Helion & Company Limited
Unit 8 Amherst Business Centre
Budbrooke Road
Warwick
CV34 5WE
England
Tel. 01926 499 619
Email: info@helion.co.uk
Website: www.helion.co.uk
Twitter: @helionbooks
https://helionbooks.wordpress.com/

Cover photograph: The fifth generation ASRAAM missile is seen being fired from a Typhoon in March 2007 against flares towed by a Mirach drone flying over the Aberporth range in Cardigan Bay, Wales. This air-to-air missile had excellent maneuverability despite the small control surfaces and lack of thrust vectoring. Lacking forward aero surfaces is unusual but appeared the wave of the future in the 1980s when the system was conceived. (MBDA by G.H. Lee)

Cover artwork: The first successful AAM was the American IR-guided Sidewinder. It was widely accepted into service around the world beginning in the late 1950s as the GAR-8 (later AIM-9B). As no aircraft to that time had been designed from the outset to employ AAM, adopting the weapons meant a modification. One type so altered was the North American Aviation F-86 Sabre made famous during combat over North Korea. This became a first-generation AAM added to a second-generation jet fighter, the missile on inboard wing pylon/launchers. The Republic of China (ROC), on Taiwan, was then confronting the mainland Peoples Republic of China (PRC) over the Formosa Straits. The air combat between these adversaries had been on largely equal basis until September 1958 when a gaggle of ROC F-86Fs, armed with GAR-8s, sought out the enemy and shot down five MiG-15s. These were the first combat 'kills' for an AAM anywhere. (Artwork by Tom Cooper)

Designed and typeset by Mach 3 Solutions (www.mach3solutions.co.uk)
Cover design Paul Hewitt, Battlefield Design (www.battlefield-design.co.uk)

ISBN: 978-1-804517-22-2

British Library Cataloguing-in-Publication Data
A catalogue record for this book is available from the British Library

We always welcome receiving book proposals from prospective authors.

CONTENTS

Note: In order to simplify the use of this book, all names, locations and geographic designations are as provided in *The Times World Atlas*, or other traditionally accepted major sources of reference, as of the time of described events.

ABBREVIATIONS AND ACRONYMS

AAA anti-aircraft artillery
AAM air-to-air missile
ABF annular blast fragmentation
AC alternating current
AESA active electronically scanned array
AEW airborne early warning
a.k.a. also known as
AM amplitude modulated
AoA angle-of-attack
ARH active radar homing
ARM anti-radiation missile
AWACS Airborne Warning and Control System
BIT built-in test
BR beam rider
BVR beyond visual range
C Centigrade
cal caliber
CCM counter-countermeasure(s)
CG center of gravity
cm centimeter(s)
CM countermeasure(s)
con-scan conical scan
CP center-of-pressure
CW continuous wave
DA direct action
deg degree(s)
DoD Department of Defense
ECM electronic countermeasure(s)
EM electromagnetic
EO electro-optical
F Fahrenheit
F^3 form-fit-function
FCS fire control system
FM frequency modulated
FOR field-of-regard
FOV field-of-view
FPA focal plane array
fps feet per second
ft foot, feet
FUR full-up round
g acceleration due to gravity, load factor
GCI ground-controlled interception
Gen, gen generation
HMD helmet-mounted display
HOBS high off-boresight
HOJ home-on-jamming
HTK hit-to-kill
HUD head-up display
Hz Hertz (cycles/sec)
IFF identification friend-or-foe
IFOV instantaneous field-of-view
IIR imaging infrared
IMU inertial measurement unit
in inch, inches
IR infrared
IRCCM infrared counter-countermeasure(s)
IRCM infrared countermeasure(s)
IRH infrared homing
IRST infrared search and track
J/S jamming-to-signal ratio
K Kelvin
kg kilogram(s)
km kilometer(s)
kN kilo-Newton(s)
kph kilometers per hour
kts knots, nautical miles per hour
lb pound(s)
lbf pounds force
LOAL lock-on after launch
LOBL lock-on before launch
LOS line-of-sight
m meter(s)
MANPADS man-portable air-defense systems
MAX maximum
MCLOS Manual Command Line-of-Sight
mi statute mile(s)
mm millimeter(s)
mph miles per hour
mps meters per second
MTBF mean time between failures
MWS missile warning system
NASA National Aeronautics and Space Administration
nm nautical mile(s)
PbS lead-sulfide
PF proximity fuze
PRF pulse repetition frequency
R radii, radius
R&D Research and Development
RAM radar absorbent material
RCS radar cross-section
RF radio frequency
ROE Rules-of-Engagement
rpm revolutions per minute
RWR radar warning receiver
SAM surface-to-air missile
SARH semi-active radar homing
sec second(s)
S/N signal-to-noise ratio
spec(s) specification(s)
TM telemetry
TV television
TVC thrust vectoring control
UAV uninhabited air vehicle(s)
USA, US United States of America
USAF United States Air Force
USSR Union of Soviet Socialist Republics
UV ultraviolet
VLRAAM very long-range air-to-air missiles
W&B weight-and-balance
WVR within visual range

PREFACE AND ACKNOWLEDGEMENTS

The advent of air-to-air missiles (AAM) brought another evolution to air combat that has always been notable for its profound changes across its history. While this AAM progress has been explored to various levels of detail over the decades, this book, spanning six volumes, will do this to an unprecedented level of detail.

This first volume begins by explaining the rationale behind AAM, how they are developed, the underlying technology, their components, and how they are employed in combat. In subsequent volumes each missile type and model within each AAM generation of systems are examined. The maturation of each weapon is explored, how it advanced the state-of-the-art, and then its deployment. The combat experience with these AAM is detailed before moving on to the next generation in a subsequent volume.

Volume 1, then, provides the background underpinning an understanding of the subjects presented in subsequent volumes where only minimal or no repeat of this material are provided. The technical descriptions of the development, manufacturing and functioning of the missiles, plus employment tactics, are meant to enhance appreciation for what each AAM contributed to the air combat milieu. It is certainly not meant to be a primer for engineers or warfighters. Likewise, the drawings and diagrams illustrating the text are by no means detailed technical renderings. Missile subjects are organized by nation, then chronologically, by significance of the system, or alphanumeric designation sequence. Within each generation, nations are organized by the magnitude of the technological contribution or number of systems developed.

In several cases the specifics of an AAM are unclear owing to remaining secrecy, confused records, dearth of research information, or intentional misinformation. Available documentation of AAM is much less than for aircraft. The combat record is also muddled and riddled with disinformation or outright propaganda. Consequently, all is presented with commentary on the reliability of the data where appropriate. Combat record 'kill' tables were principally drawn from the ACIG.org website database, but augmented with specific information from other sources, especially in identifying the models of AAM for each kill. There are many other summaries of kills available for each conflict with slightly different results recorded. The author and publisher chose to use ACIG.org for consistency and because it was suitable in displaying the scope of AAM use. There are no indisputable sources.

The reader is expected to come to these volumes with a fundamental understanding of how air vehicles fly plus a basic background on air combat and the evolution of aviation. Past tense is used throughout for the distant future reader except for Volume 1's technical foundation. Apart from the sources given in the bibliography at the back of these volumes, hundreds of other books, articles, papers, online references, and personal communications provided information for this book, far too numerous to list in detail.

The author is indebted to the following for their kind assistance in this project: the folk at the Air Force Armament Museum, Tom Cooper, Ron Feldman, Jeannine Geiger and Dr Michael Doidge at the Edwards Air Force Base Air Force Test Center History Office, Albert Grandolini, Tony Landis, Larry Milberry, Jay Miller, Sean O'Connor, Jim Rotramel, Kim Sheldon and Bob Thomas at the National Naval Aviation Museum, Will McLaughlin of the National Museum of the United States Air Force, Patricia Sauers and the Naval Base Ventura County archives, Dr Bob Smith and the folks at the China Lake Museum, and Mark Swaney.

Bill Norton (nortowj1@gmail.com)
Tehachapi, California, 2025

1

INTRODUCTION

> Tools, or weapons, if only the right ones can be discovered, form 99 percent of victory ... Strategy, command, leadership, courage, discipline, supply, organization and all the moral and physical paraphernalia of war are nothing to a high superiority of weapons – at most they go to form the one percent which make the whole possible.
>
> J.F.C. Fuller

> The Air Force's inflated rhetoric on missile efficacy, based on misassumptions and technological exuberance, was punctuated by the failure of missiles in actual combat.
>
> Lieutenant Colonel Steven A. Fino, USAF

These quotes emphasize both the essential character of weapons technology in modern warfare, especially that waged in the air, and the risk of becoming too enamored and encumbered by those same weapons. Striving for success in air combat has been motivation for many technological achievements in weapons development. For air-to-air engagements, the desire to shoot down opposition aircraft at greater range and with greater effectiveness has propelled air-to-air missile technology across 80 years of progress and combat. Paced by available technology, the ever-changing characteristics of the carrying aircraft, and lessons from air battles, these weapons have evolved through five generations and dictated progressively altered tactics. It has been a history peppered with failed or marginally performing systems, great wealth expended fruitlessly, misperceptions by the public and warfighter alike, and mixed combat results. Yet, the air-to-air guided missile remains a weapon dominating aerial engagements.

Fundamentally the AAM (Figure 1) is a guided projectile (generically a 'round') launched from an attacking aircraft towards a target aircraft. The missile, propelled by a motor, possesses a seeker head to sense the target via one of several means. The seeker detects and 'locks' onto the target, tracking its movements. The most common AAM have been infrared homing (IRH),[1] semi-active radar homing (SARH) and active radar homing (ARH). The guidance system uses the seeker data to determine position of the target with respect to the missile and directs the AAM to closely approach or strike the opposition aircraft. An autopilot steers the round, given the guidance input, via a flight control system employing, most commonly, aerodynamic surfaces akin to those of an airplane. A fuze detects the close approach to the target and detonates a warhead with intent to destroy or fatally damage it. A great variety of carriage and launch methods, propulsion systems, guidance means, fuzing and warhead types are to be found in the more than 100 AAM types, and many more models, developed over 80 years. They have roughly broken down into short-range and medium- to long-range missiles. The short-range types are used for close-in 'dogfighting' engagements within-visual range (WVR, typically about 3mi/5km),[2] guiding principally on infrared (IR) 'heat' emissions of the target aircraft. The medium- and long-range types are used for beyond visual range (BVR) engagements and guiding principally on reflected radar energy.

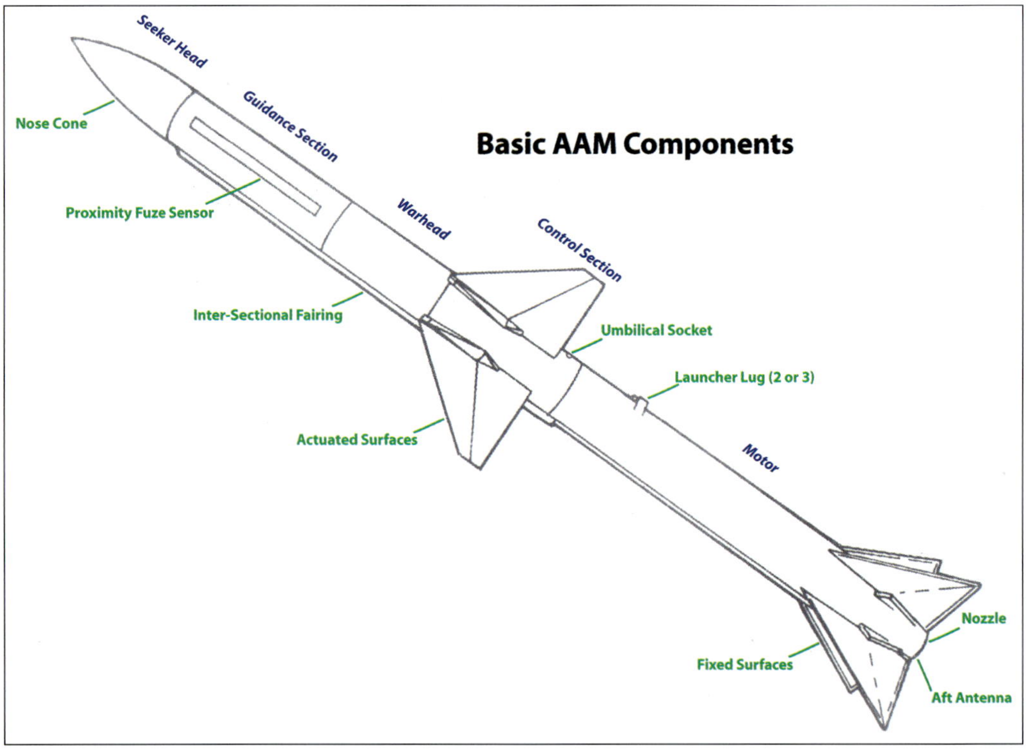

Figure 1: The basic modular components of an AAM are depicted. There is much variation in such features. (Drawing by author)

Fundamentals

The early decades of air combat required one aircraft to closely approach another in order to bring gunfire to bear in an effort to drive away or shoot down the opponent. This typically necessitated similar or superior airspeed and acceleration capabilities to overtake the opponent. The range and effectiveness of this attack means increased with higher caliber machine guns and cannon, greater muzzle velocity, rate of fire, number of weapons aboard the attacker, ammunition quantity, and also gunsight improvements. The attacking pilot had to consider range, lead angle, and dispersal pattern of the rounds sent towards the target that was likely maneuvering to evade. (Lead is to aim and fire ahead of a moving target such that the rounds and the target arrive at the same point in space at the same instant. This is also called a deflection shot.) By the time of the Korean War in the early 1950s, successful jet combat gunfire was dependent on a ranging radar, computing gunsight, gun

An F-20 prototype fighter is seen firing a short-range AIM-9 'heat-seeker' from a wingtip rail launcher. This Sidewinder AAM served in most air forces around the world for decades and figured prominently in numerous conflicts. The motor exhaust has little smoke, and the Mach or Shock Diamonds are an artifact of the supersonic flow. The gases in the plume expand to ambient air pressure and the flow is compressed repeatedly into shock waves made visible by the continued burning of combustibles in the exhaust. (Author's collection)

Above: An AIM-54A is fired off a US Navy F-14A during a test. This long-range radar-guided Phoenix missile represented a tremendous step forward in AAM when introduced during the 1970s. The Tomcat and Phoenix fighter/missile combination was used to good effect by the Iranians during their war with Iraq throughout the 1980s. (National Naval Aviation Museum)

Right: A Phoenix missile destroys a QF-9J Cougar target drone during a test on 26 November 1974. The missile appears to have detonated close to the target drone after the fuze detected the aircraft and set off the warhead. The blast and fragments have pierced the fuel cells and ignited the fumes as well as destroying the engine and damaging the airframe. (National Naval Aviation Museum)

heaters and firing circuits. These and gunsight variabilities meant that experience counted for much in being able to hit a target, yet combat experience is a persistently vanishing quality.

When the quarry had a gunner of its own to discourage attackers, the predator sought to engage from outside the range or field-of-fire of those weapons. A cannon (assuming 20mm caliber), with its fuzed/exploding shell, typically has an effective range of approximately 3,000ft (900m). The number of rounds and rate of fire typically permits only a couple of trigger squeezes, or a few seconds of continuous firing, before ammunition is exhausted. When closing on a bomber, the fighter pilots might face a rear turret firing with rounds having the additive speed of the bomber and so out-ranged the attacker by 300–3,000ft (90–900m). Table 1 shows that Soviet interceptor pilots were aware of the vulnerable ranges and that there was good probability of taking hits during an encounter. Especially naval aviators could expect to face a bomber more like a fighter and so with speed more closely matching their own. Opportunities to achieve a deflection run from abeam would be few and head-on attacks would leave precious little time to sight and fire given the high combined closing speed. The Americans explored articulated gun installations with gun-laying radar that did not rely so much on setting up gunfire runs. These proved heavy and awkward, never leaving the trials stage. In defense of a gunfire attack the targeted pilot can perform many hard maneuvers to spoil the attacker's ability to keep the aircraft in the gunsight reticle. (For a computing gunsight the lead is provided automatically.) This 'jinking' increases the deflection to a much more difficult 'snapshot' by complicating the shooter's ability to gain a favorable position.

In the early 1950s some air arms used unguided rockets air-to-air. For the Americans this was the 2.75-inch Folding-Fin Aircraft Rockets (FFAR) with a nominal range of 13,500ft (5,115m). These spread out with intent at least one would strike the target, principally a large bomber, to detonate the approximately 6lb (2.7kg) warhead. An adverse characteristic of the rockets was their tendency to 'weathercock' into wind shears and gusts. Training scores, or lack thereof, did not indicate high probability of success.[3] These rockets were a gap-filler until guided missiles could be introduced. Fortunately, the wait was not long.

The air-to-air guided missile held the promise of greater range and less effort in bringing deadly fire to bear. In the case of the Soviet fighter pilots vulnerable to bomber tail gun fire (Table 1), for example, they were to fire AAM outside the vulnerable range, if able. In general, AAM shots would not require complex maneuvering or time-consuming closure to proximity suitable for a gun solution. They might also demand less skill and so reduce air combat maneuvering (ACM) training requirements and cost. Jinking became largely moot because the effective field-of-fire for an AAM is greater than for gunfire and such small maneuvers will likely not defeat the missile. There was also the potential for simplified night engagements as locating the adversary and closing to make visual contact in the dark would not be necessary. However, the potential for the attacking pilot to be momentarily blinded by the glare of the AAM's motor at night was the same or worse than by muzzle flashes and tracers. Particularly with a high-energy rocket motor, the missile would fly at much greater speed than the quarry, reducing the potential for

The American .50cal machine guns were a significant weapon in air combat during the Second World War and carrying on after until they just did not have the required penetrating momentum 'punch' against more ruggedly built high-speed jet aircraft. As seen here, as an armorer attends to the trio of guns on one side of the nose of an F-86D, these were substantial pieces of hardware, each weighing 65lb (30kg), with large rounds. The six guns were supplied with a total 1,800 rounds. (National Museum of the United States Air Force)

The twin 30mm cannon assembly of the Mirage IIICJ is displayed in front of an Israeli jet. Obviously weighing hundreds of pounds, it was designed to be readily removed from the aircraft for servicing. The Mirage IIICJ carried 125 shells that were exhausted in seven seconds of firing. The Israelis had Dassault install the cannon on the Mirage III that initially lacked guns. (Author's collection)

An interim solution, bridging the cannon and the AAM, were unguided rockets intended to bring down slow-moving bombers. Both the US Air Force and Navy adopted such weapons, plus a few foreign services. Here, an F-86D (top) shoots 2.75-inch projectiles from an extendable belly tray and an F-89B (bottom) from wingtip pods. Spreading out during their ballistic flight, perhaps one projectile would hit a large target. (National Museum of the United States Air Force and National Archives)

its pilot to evade. Conceivably, even the pilot of an inferior aircraft could defeat a superior adversary. However, with the early generation of missiles much still remained dependent on attacker aircraft performance, pilot skill, and the ability to intercept the target to advantageous conditions (closing speed and angle of intercept). However, unlike guns and cannon, missiles could and did evolve to reduce these factors for improved effectiveness.

Table 1: Percentage Probability of MiG-21 being Hit on One Pass by Bomber Tail Gun[4]

Pullout Range (nm)	**Closing Speed (kts)**	**B-52 or B-58**				**B-47**			
		55	80	110	160	55	80	110	160
0.43	**Probability of Hit (%)**	42	34	27	10	25	19	15	11
0.33		54	44	36	26	34	26	21	16
0.25		68	48	48	35	45	36	27	21

Fighter pilots had just adapted (some reluctantly) to cannon with fewer (albeit exploding) shells and a lower rate of fire than the earlier machine guns. It was now being proposed that they be satisfied with two to four missiles. (Without aerial refueling support a jet fighter seldom has the fuel to perform more than one or possibly two engagements.) The prevailing suspicion was that guided air-to-air missiles would be too expensive, would be too large, heavy and draggy, could not be maintained in the field, demanded a fire control system (FCS) that added to employment complexity,[5] and would be simply impractical for real-world air combat. When introduced, and with many of these preconceptions shown to be false, there was still hesitancy.

Even the earliest missiles flew several times the speed of the target aircraft. Yet, given the range, flight time was very long, and the pilot had to wait a comparatively long period to see the result of the shot. The AAM can possess a time-to-impact (TTI) of generally no more than 5–15sec for short-range, 15–25sec for medium-range and 30–40sec for long-range. Shooting a missile and awaiting the result as a spectator did not sit well with the first jet fighter pilots adapting to the weapons. The reliability of the initial generation of AAM did not inspire confidence in success, some seeing them as a 'wish-'em-dead' weapons – push the firing button and hope for the best. Especially when fighters were produced without cannon and equipped only with missiles did the 'tire meet the road' and deficiencies come to the fore. Countermeasures to confound AAM tracking also evolved. These factors prompted firing salvos of missiles but also continued improvements (and the cannon restored) until the AAM met the promise. By the 1990s the AAM had become the principal air-to-air weapon, accounting for 90 percent of kills. By the twenty-first century and fifth generation of AAM the probability of the missile functioning properly and knocking down the target on the first shot had risen tremendously. It then became a matter of evolving tactics and combat circumstance, plus rules of engagement dictating when and how the AAM would be employed. System knowledge, cockpit management, information filtering, some physical attributes, and professionalism (a relative term) then became the valued traits of the fighter pilot.

Development of AAM began during the Second World War, driven by the imperatives of that conflict and the rapid advancement of armaments and aircraft systems. These factors were especially true in the immediate post-war period as jet aircraft entered service in increasing numbers. These brought higher airspeeds that necessitated tougher construction. The AAM had to be several times faster and the warhead more lethal. At very high speed, even after motor burnout, the missile can continue to maneuver and close on the target even as energy bleeds off owing to friction from passage through the air. Generally, the faster the missile the better, though turns grow in radius or g-level (load factor) with airspeed.[6] An aircraft flying 400kts and turning at 9g will describe a turn radius of 0.3mi (0.5km) whereas the pursing missile at 2.5 Mach turning at 35g will have a 1.3mi (2.1km) turn radius.[7] The missile could well overshoot and have to correct back, bleeding energy, assuming it had not lost lock. All this also demands highly efficient fight controls and great airframe strength.

Above about 4 Mach the kinetic heating of the missile from flight through the atmosphere can require special design measures that entail cost, weight and complexity. Lacking these, the heated airframe can suffer degraded load-bearing capability.[8] For the first

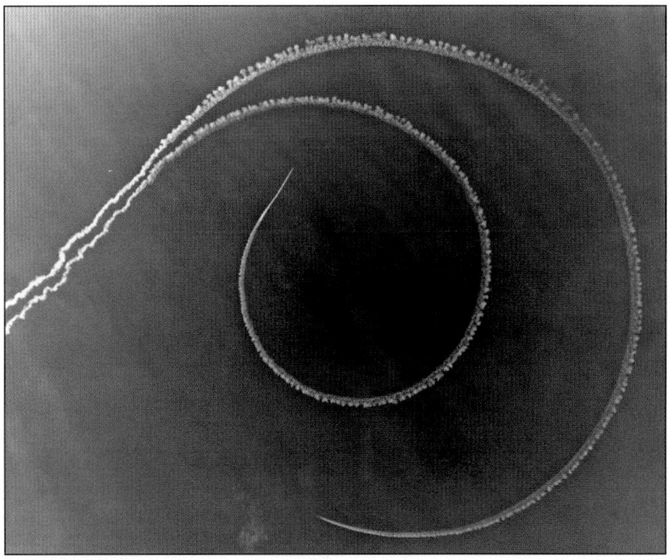

The famous photo of an F-16 turning inside an F-4 illustrates, via the contrails, the advantages of a highly maneuverable fighter in gaining an air-to-air combat advantage over a less-advanced adversary – at least in ability to get on their tail for a gun or infrared homing (IRH) missile shot. It also shows relative turn radii regardless of g pulled that may be just as important. (US Air Force Test Center History Office)

production AAM, the missiles' sensors and electronics were greatly limited by the technology of the period. This made for an extended development period and the weapons suffered low reliability and effectiveness. Military weapon system development programs were then driving some basic technological maturation such as electronics miniaturization and 'ruggedization' while technologies specific to weapons, such as sensors, motors and warheads were advanced in government laboratories either as ongoing research or as a program to develop a specific capability (research and development, R&D). With the need to introduce progressively improved AAM, in answering new capabilities of potential adversary systems, a series or generations of missiles emerged around the world. As in all such weapon programs, some enjoyed superiority, in country of origin or abroad, until a responding counter capability was fielded. This demanded development of the next, more capable generation of AAM in an endless evolutionary cycle.

The added weight and drag of missiles degrades aircraft performance and maneuverability. Given missile weight, volume, and other fighter mission requirements, it was recognized from the first that the aircraft would likely carry only two to four AAM and so the number of potential engagements were few. Additionally, being far more complex than guns and cannon, the failure rate of the AAM was expected to be greater. However, even cannon have a limited number of shells and a few pulls of the trigger usually depletes the magazine. It was assessed as wise to have both AAM and cannon available to the fighter pilot and so potentially extend combat duration. Regardless, the idea of an all-missile fighter had appeal from several perspectives. The desire to fly faster to rapidly engage, especially with nuclear-armed bombers as targets, meant supersonic fighters and missile launches from a distance. The age of the turning dogfight seemed to be fading away. To get the most from these fighters, their weight was reduced for favorable thrust:weight ratio and so wing size was likewise reduced to also cut mass and aerodynamic drag. With the expectation that the opposition would follow the same trend, the usefulness of a heavy gun in high-speed fights became minimal and they were eliminated. However, with apparent success in developing practical AAM, the design 'pendulum' swung too far in one direction. The high wing load aircraft (aircraft gross weight divided by wing area) were less maneuverable in a turning tangle with earlier generation low wing load fighters that retained guns. With the performance of the first generation of AAM being uninspiring, the gun-less, high wing load fighters suffered some important disadvantages. It was found that even a missile-only machine had to be able to perform reasonably tight turns to enter a good shot window. Consequently, the gun was restored and the pendulum swung back towards what was hoped (at least in the near-term) would be a 'happy medium.'

With the potential for any number of combat scenarios, it became advantageous to carry WVR 'dogfighting' AAM and BVR missile on the same fighter. The Soviets particularly favored this approach by, for decades, developing SARH and IRH models of the same missile type to save on development and support costs. The best approach has generally been to fire the IRH rounds first so that, if the SARH is sent first, the heat-seeker does not home on the intense motor signature of the initial round. This, however, works counter to what could be envisioned as a likely engagement scenario with the SARH launched BVR and the fighter continuing to drive in to the target, radar-painting the opposition machine, until WVR

The American F4H Phantom (later F-4) was conceived as an all-missile fighter without gun. It was given a heavy load of AAM as one configuration option for the fighter-bomber as suggested by this image – four AIM-9 on pylon-mounted rails and four AIM-7, conformal/semi-submerged in the fuselage-wing blend, matched to a powerful and capable radar. This supported both WVR and BVR engagements of an adversary aircraft. (Ra'anan Weiss collection)

for the IRH round. In any event, having two types of AAM inbound near-simultaneously places great demands on the targeted aircrew's ability to evade and decoy the threats with countermeasures.

For aircraft carrier-based aircraft, bring-back load capability can mean jettisoning stores prior to landing back aboard ship. This is because of limitations in landing gross weight to ensure against structural overload of the aircraft landing gear and the carrier deck. Jettisoning costly and scarce AAM into the sea was undesirable and only overcome with aircraft design advancements.

Characteristics

There are several basic characteristics of AAM affecting air combat, or these in part dictating how engagements are executed:

- Range of the missile and/or associated launch aircraft sensors (minimum and maximum)
- Sighting capability aboard the shooter aircraft
- Engagement angle (aspect) capability of the AAM with respect to the target and so maneuvering required of the shooter for a suitable shot solution
- Energy lost by the shooter during maneuver, or required loiter, placing the aircraft at risk of engagement by adversary aircraft or air-defense systems
- Onboard sensor-to-missile interface
- Speed and expected flight time of the missile requiring the shooter to remain engaged (depending upon the nature of the AAM) and potentially allowing the quarry to evade and the shooter to be engaged by another adversary
- Circumstances affecting the potential for successful tracking of the target once the missile is in free flight to include enemy countermeasures
- Circumstances affecting the potential for successful warhead fuzing and detonation
- Potential that the warhead causes lethal damage or at least forces the adversary to disengage and withdraw

It is advantageous to possess a missile that can be fired from wide aspect angles between the shooter and the target ('angle off the tail') and, from the missile's perspective, off-boresight (beyond the longitudinal axis of the round). See Figure 2 that introduces the concept of line-of-sight (LOS). For the AAM, this equates to a good field-of-view (FOV) or the span of degrees that is 'seen' by the sensor, like the vision of the human eyeball without turning the eye. The sensor is typically gimbaled to look around to its mechanical limits. This is the field-of-regard (FOR), similar to the limits of human sight by turning the eyes. (Turning one's head is analogous to turning the aircraft on which the missile/seeker is mounted.) These relationships are shown in Figure 3. Good FOV and FOR can reduce required maneuvering by the shooter to obtain a firing solution from a few minutes to tens of seconds while also reducing the potential the enemy can evade and the shooter engaged by other adversary aircraft. Generations of missile development have sought the All Aspect Capability (ALASCA) AAM with the advantages illustrated in Figure 4.

The missile's rocket motor and the AAM's aerodynamic drag will largely determine peak flight speed, setting aside energy-draining maneuvering and air density at the operating altitude. Range and endurance of the flight is then a function of the duration of burn and the airspeed achieved. At propellant burnout and end of thrust the AAM's speed begins to decay immediately (unless descending) owing to aerodynamic drag, maneuvering and density of the air the round is passing through. The deceleration means a drop in

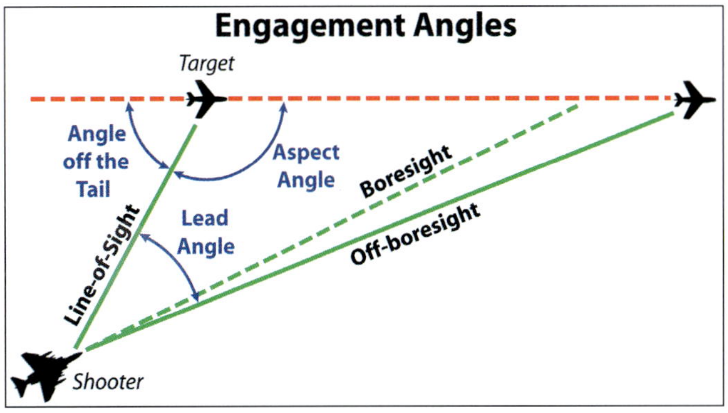

Figure 2: Depicted are the relationships between the engaging aircraft and adversary defining aspect angle plus off-boresight. (Drawing by author)

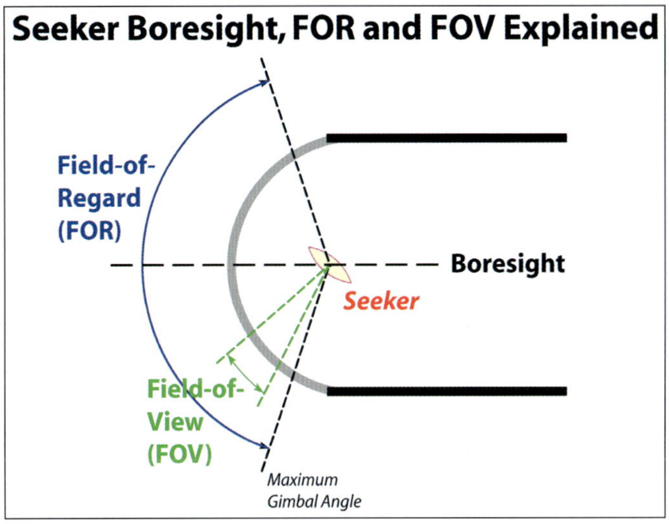

Figure 3: This simplified cutaway of the forward end of an AAM (seeker head) illustrates the concepts of boresight, field-of-view, field-of-regard, and maximum gimbal angle of the seeker. This is a two-dimensional representation of a three-dimensional physical reality. (Drawing by author)

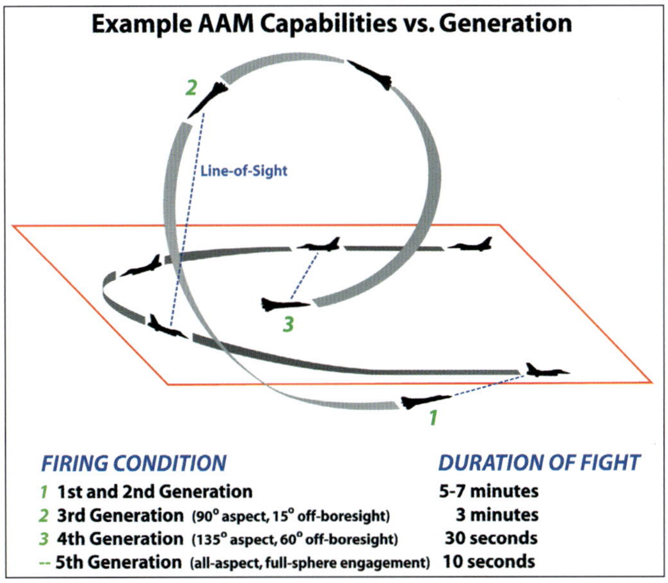

Figure 4: This drawing illustrates the conceptual maneuvering and firing line-of-sight (dotted lines) by the engaging aircraft pilot to achieve an AAM shot solution possible with the various generations of missiles and the notional time for the engagements. (Drawing by author)

maneuver potential as control effectiveness (Chapter 3) diminishes until the AAM falls away from the intended flightpath owing to gravity. This can affect the ability to complete the endgame intercept of a maneuvering target to within lethal radius of the warhead. The range numbers presented in these volumes for missile type/model characteristics are usually optimal 'brochure' numbers or values expected for 'typical' combat scenarios. These usually imply co-equal altitude engagements. A target above requires the missile to climb and slant range would be less than for level flight which is less than in a descent.[9]

With a narrow 'squint' FOV the initial 'dogfighting' heat-seekers were only suitable for visual-range and tail-chase engagements (small angle off the tail) in which 3mi (5km) range (WVR, at low altitudes) was practical and with a lightweight AAM (around 220lb/100kg or less). However, aspect angle can also mean extended range as the closing speeds are much greater in the forward hemisphere where airspeed of the AAM and the target are summed to some extent but subtracted in the rear quarter (Figures 5 and 6). Figure 7 shows this effect of flight speed of the shot plus the flight speed of the missile decreasing range – aerodynamic drag increasing with higher speed. If the target is pulling away the effect is further amplified. Missile potential range can be reduced as much as 50 percent or more.

Although operating at higher altitude than the motor nozzle is optimized for (see later) can compromise thrust, the more rapid decrease in aerodynamic drag with the lower density of air generally means greater range, as shown in Figure 8. Yet, aerodynamic control is also degraded in lower density air at altitude, affecting maneuverability needed for endgame closure on the target. Much maneuvering also drains energy and so airspeed, further degrading range. At 50,000ft (15,000m) and head-on the range of an example heat-seeker, 3mi in a tail-chase mentioned earlier, could be around 15mi (25km) provided the seeker has the ability to lock onto the target in that aspect and range. Clearly, range is dependent on many factors of the engagement and may not be the decisive factor in achieving a 'kill.'

In scenarios during which the engaging aircraft faces a numerically superior foe it can be advantageous to reduce the number of enemy aircraft using long-range, possibly radar-guided, AAM before closing for 'dogfighting' in which short-range, possibly 'heat-seeking', AAM would be employed. Basic 'desirements' for the dogfighting missile is a small minimum range, rapid launch and arming, good off-boresight capability, and high-g turning. However, carrying two types of AAM was not initially practical. The radar-guided models are bulkier

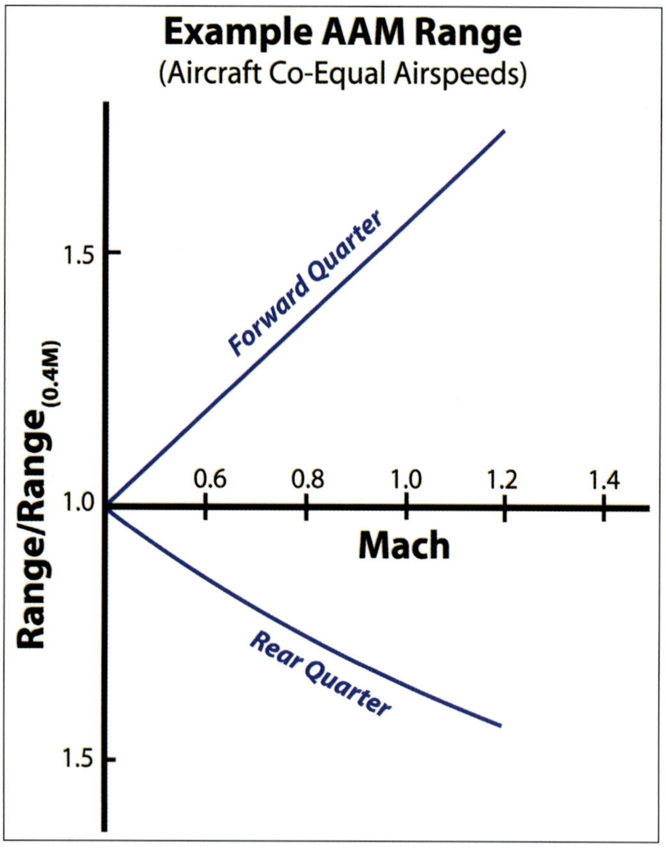

Figure 5: Depicted is the effect of target/shooter co-equal airspeed on missile range, normalized to 0.4 Mach, for rear-quarter and forward quarter shots. Head-on shots are clearly achievable at a greater distance at all airspeeds, though likely beyond visual range. (Drawing by author)

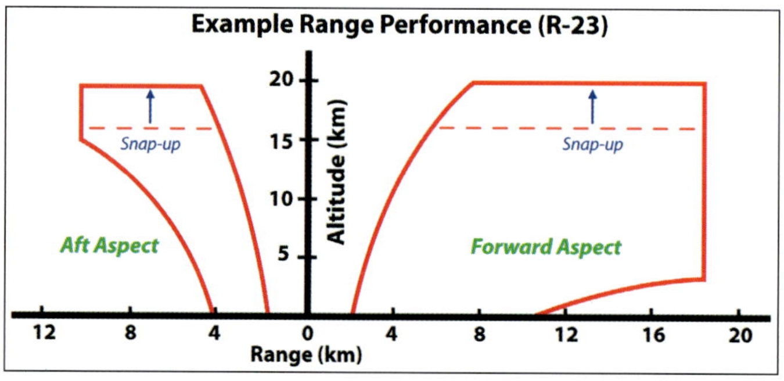

Figure 6: Shown is the basic range performance of the Soviet R-23 missile in front and rear aspects as a function of altitude. This aptly illustrates the better maximum range performance at higher altitudes than lower levels, and forward versus aft aspect. Minimum range, discussed later, is also affected. Included is an indication of the ability to engage targets above the launch aircraft (snap-up), this applicable at all lower altitudes as well. The R-23 system had, in actuality, no snap-down capability. (Drawing by author)

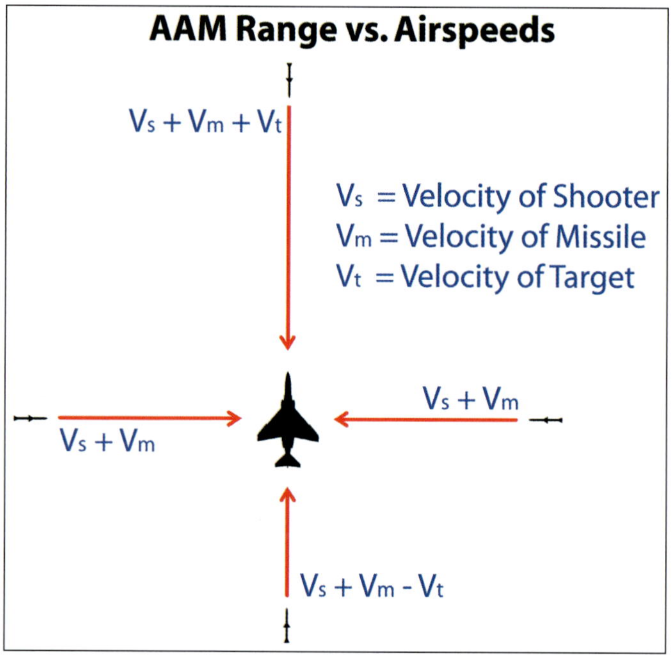

Figure 7: Another look at range as a function of shooter and target airspeeds (velocities, V) is shown at the four basic quarters of engagement. Fundamentally the greatest range is in the forward quarter, provided the seeker is effective from that aspect. The shortest range is a rear aspect shot where the seeker may be the most effective. The beam shot can be the most challenging for the seeker yet offers respectable range. (Drawing by author)

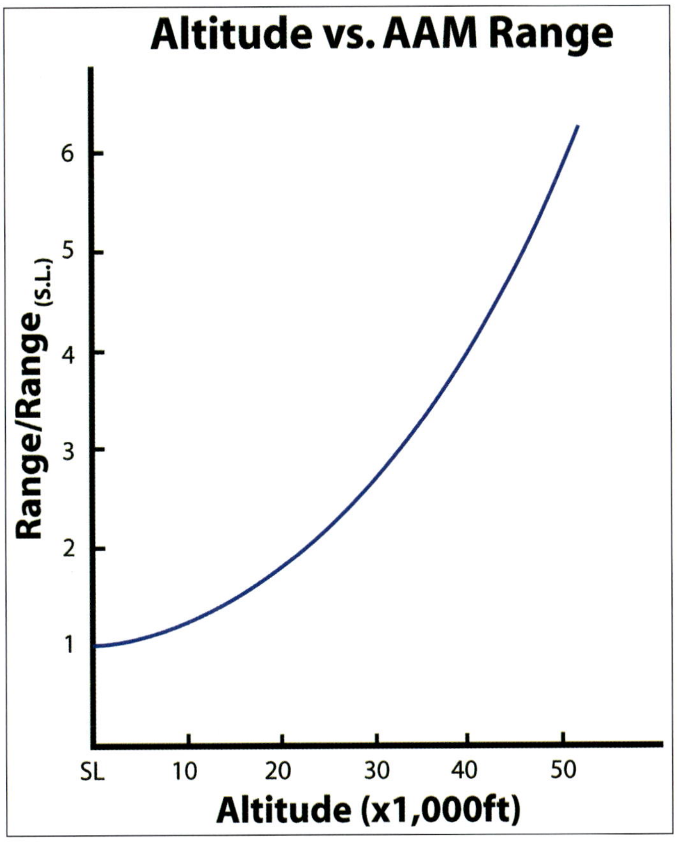

Figure 8: The relationship of altitude on missile range is depicted, normalized to 1,000ft above sea level (S.L.). (Drawing by author)

and heavier, displacing other stores, and more dependent on aircraft systems. The heat-seekers can be employed by aircraft not equipped with radar or with radar unsuitable for operation of radar-guided AAM. Additionally, the reliability of the initial generations of aircraft radar were relatively low and so the dogfighting AAM offered more probability of a successful engagement.

Chasing Capabilities

By the late twentieth century wars had become brief affairs, on the order of weeks. This did not permit time to adapt weapons to the unique circumstances of the combat. The saying that you go to war with the air force you have is quite *a propos*. Nations and air arms need to anticipate needs by studying the capabilities of potential adversaries and threats in the evolving aerial battlefield. The insight gleaned informs requirements for upgraded or new weapons. Most air arms have some personnel devoted to studying and characterizing the weapons of potential adversaries so that superior systems can be sought, or tactics and countermeasures adjusted to make the most of what is in the inventory. Seldom is it possible to examine such adversary weapons directly, much less test them and possibly extract some useful technology, although this has occurred occasionally over the 80 years of AAM history.

The development and maturation of AAM can be an element of an arms race. This is either replacing an obsolescent (worn out or unsustainable) weapon, offering no advantage over superior enemy munitions, or responding to a superior system introduced by potential adversaries to gain an advantage. Reducing long-term operating costs and creating weapons that can be used by multiple military services ('joint' programs within a nation or an alliance of nations) can be other motivations. As nations other than the superpowers began developing, manufacturing and marketing these products, AAM also became an element of international arms sales competition. Certainly, within international defense alliances there is a desire for all to operate identical or similar systems for commonality and ready support in wartime. However, nations seek self-sufficiency and ability to field weapons more suited to their particular circumstances. This can be important in situations where an embargo may be imposed that will deny operators support such as the replenishment of combat expenditures, supply of repair parts and assemblies, or even notification of revealed deficiencies. Local production also offers the opportunity to control features such as lower susceptibility to countermeasures employed by potential adversaries. These might also hold national secrets and ensure an advantage through lower potential for disclosure to enemies or competitors.

Israeli personnel were transported to the wreckage of a downed Egyptian MiG-21 somewhere in the Sinai desert, during March 1969, to examine a partially intact R-3S missile. Little can be learned by a visual examination in the field such as this, and it can be assumed the round, with motor and warhead as hazardous material, was transported back to a laboratory. Being able to explore adversary weapons in this manner is rare. (Albert Grandolini collection)

Aircraft Integration

The aircraft usually interfaces with the missile in various ways. These include electrical power supply and possibly coolant transfer prior to launch. Fire control system data may also be transferred, requiring compatible computer data formats. More advanced systems may also have transmitted datalinks requiring transceivers and antennas. All of these require, to some extent, monitoring and control by the operator and aircraft systems to permit, at a minimum, a launch command and possibly a motor ignition electrical charge. Common are also signals indicating that the missile sensor has acquired the target, and a successful launch/engagement is then possible should the pilot choose to fire.

It is seldom that a new aircraft and missile type are developed simultaneously with intent that they be fielded together. More common is that the new AAM must be compatible with existing aircraft, or new aircraft compatible with existing missiles. The ideal is that a new AAM will match form-fit-function (F^3) of prior weapons such that the intended warplanes need not be altered in any way to accommodate the system. Particularly on fighter-type aircraft, volume and power margins are at a premium and any changes can cause a cascade of other adjustments to established systems and equipment, introducing the potential for unintended adverse consequences, delays, and elevated cost. Especially adding another control head and/or display to an already crowded cockpit and instrument panel can be very disruptive. All such changes are carefully reviewed by a Cockpit Working Group or similar body that includes representatives of the operating commands, depots, and developers. Likewise, for changes to existing displays to show new symbology and new functions. Aircrew would have input on the appearance and functionality of such equipment (human-machine interface, HMI, or pilot-vehicle interface, PVI). The requirement to upgrade radar, displays, computers, and more to integrate, say, a fourth-generation SARH missile onto a fleet of third-generation jet fighters must be weighed against the cost of buying new aircraft. All such changes also bring training and logistics implications. Hence, the modifications could exceed the original purchase price of the jet. The fielding delay imposed by the years-long fleet modification ('mod') period must similarly be weighed against the projected benefit in additional defense capability.

However, F^3 is seldom the case, especially with highly integrated and software-intensive twenty-first century aircraft and missiles containing millions of lines of computer code and with subsystems that must 'talk' to each other for suitable interoperability. Integrating a new weapon onto such aircraft requires software recoding and then verification checks that this operates as expected and without introducing unexpected results. Even upgrades in the missile, initially accomplished primarily with electronics changes and later via successive software 'builds', can have similar implications for aircraft/missile compatibility.

Hardware changes require aircraft modifications that must be designed and tested. Modification kits are assembled and provided to the mechanics team undertaking the task. The 'mods' typically require several days to several weeks (or possibly months) to complete and are done progressively such that the entire fleet is not grounded for this work. This also means that operational units have a mix of aircraft capable of firing different weapons for a time, imposing scheduling, support, and training complications. Even uploading new software code can be fraught and must be done with care to avoid inadvertently affecting some other function adversely. Given that the most modern AAM are software-intensive with considerable communication between aircraft and missile, the requirement for aircraft coding changes becomes inevitable. This can also set apart aircraft intended to operate with the new missile. An alternative is to have the new missile mimic the interface behavior of the old to alleviate these integration concerns, preferably without degrading the AAM' capabilities. While the much-maligned phrase 'software don't break!' has some truth, the code may be so complex that it does not function as expected and troubleshooting/remedy can be prolonged.

Aeromechanics

It is desirable that carriage of the AAM has minimum impact on aircraft performance and handling qualities or flying qualities,[10] even in download configurations (some but not all missiles fired or jettisoned, possibly leading to an asymmetrical loading across the aircraft span). Conformal mounting with the missile partially submerged in a depression within the outer surface (outer moldline), has been used to reduce drag, though ensuring seeker head detection of the target prior to launch is challenging. A missile mounted at a wingtip was actually found to have favorable effect on the drag-producing wingtip vortex resulting from high-pressure air below the wing spilling around the tip to the low-pressure zone on

Epitomizing an aircraft not initially designed for AAM employment is this Italian F-104S whose pretty lines have been ravaged by added installation of pylons, launchers and missiles, some with skoosh ground clearance should the aircraft experience a hard landing greatly compressing the landing gear shock struts or suffering a flat tire. It carries a collection of American Sidewinder and Sparrow AAM that, no doubt, severely degraded aircraft performance despite adding combat capability. (Albert Grandolini collection)

top. A weapons bay which contains the AAM within the airframe is the ideal from an aircraft drag and radar reflectivity perspective, but also most challenging from complexity, weight, and cost perspective, plus complicating weapon loading. Target detection prior to launch, again, needs to be addressed.

Separation

The AAM is mounted on the aircraft in a manner that allows its sensor to detect the target (if so-designed) and for it to depart the aircraft safely and effectively when fired. The mount or pylon typically stands the weapon off from the aircraft outer moldline and allows ready loading of the round by ground personnel. The mount includes a missile launcher as an integral component or an attachment, potentially with an interface adapter.

Some missiles must accelerate after motor ignition to a stable flight speed (beyond that of the firing aircraft), where the control surfaces are suitably effective, and so runs along a short rail as an element of the launcher (Figure 9). This also serves to retain and restrain the round during all other aircraft operations that risk 'flinging' the store away. The rail may be like a C-channel with a lug like a bolt head or wingnut projecting from the missile running within the channel. A launch rail like a railroad rail requires lugs wrapping around both sides. One historical difficulty with rail launch is that external forces and inertia are acting on the missile from the moment it moves down the rail and may impose a combination of vertical and lateral motion. This can cause the lugs to gouge the rail, especially with the last lug alone moving on the rail, and the missile is diverging from the axis of the launcher. This is called 'tip-off.'

Some AAM are suitably stable at the aircraft flight speed and can be simply dropped with motor ignition immediately after a safe separation distance from the aircraft. In some cases it is necessary to eject the round forcefully through an adverse flow field around the aircraft that risks upsetting the missile such that it cannot be controlled or risks striking the jet. The missile must still be held firmly to the launcher by mounting lugs until released at ejection. This includes uplocks that must be released at launch and swaybraces with snubbers which are adjusted to firmly press against the missile when mounted to prevent lateral motion during aircraft operation. The ejection is done via pistons within the mount that are operated electrically, pneumatically or hydraulically. The pressurized gas for a pneumatic mechanism can be stored in an accumulator or generated via small pyrotechnic charges. There are typically two ejectors, and

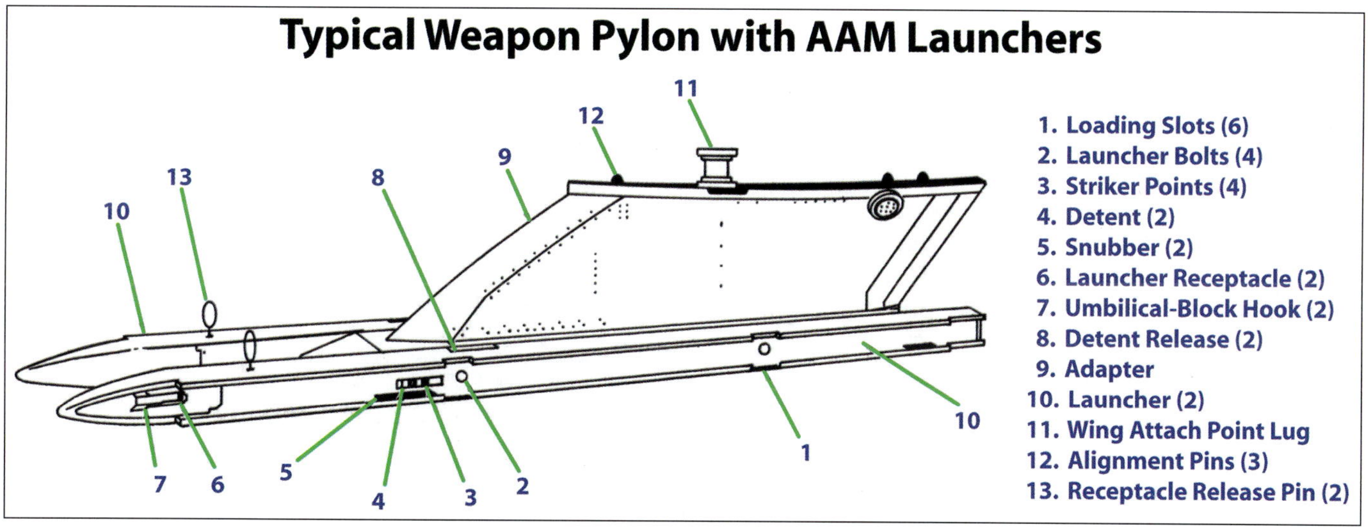

Figure 9: Shown are elements of a typical AAM weapon pylon for underwing mounting. (USAF, Modified by author)

A Meteor missile is ejected from a Rafael fighter in a safe separation flight test, an F-16 safety chase aircraft seen beyond. The two long-stroke ejectors with springs are visible in an image distorted by the wide-angle separation cameras mounted externally on the test aircraft. (Dassault Aviation)

these may operate at different rates and stoke length to send the AAM away at an attitude and induced motion for a safe separation as determined by flight tests. A slight nose-down separation is typical. Springs may be included to pull the ejectors back to a stored position. Ejection is normally felt or heard as a 'thump', giving the aircrew an indication that the store has separated even if other indications are lacking or faulty.

The introduction of radar reflection reduction low-observable or 'stealth' design principles in the late twentieth century meant that stores were carried inside weapons bays. (The Americans originally carried AAM internally on early jets to reduce drag given the comparatively modest thrust of the initial turbojet engines.) To be fired, the missiles need to be extended beyond the outer moldline of the aircraft via an articulated mount, typically a pantograph 'trapeze' mechanism. This might also be necessary for target acquisition by the seeker head unless firing a missile that acquires the target after launch. (Producing a stealthy AAM has not been sought owing to their flight being evident by various means, especially presence of the launch aircraft, electromagnetic [EM] emissions, and smoke.)

The separation of the missile from the launch aircraft must ensure dynamics do not cause a loss of lock on the target, impact with the aircraft, or contact with other stores. Firing with the aircraft in a turn can cause the missile, likely not yet guiding, to initially 'weathervane' or align with airflow around the aircraft that is not along the missile centerline. This can potentially cause the seeker to lose lock before flight controls take over in response to guidance inputs. If the motion imparted to the missile at launch (pitch, roll and yaw) exceeds the power of the control system to maintain dynamic stability, the round will depart controlled flight and be lost. Launch of the AAM can also induce an abrupt 10–20g to the round – usually longitudinal and possibly vertical – and it must be designed to withstand this acceleration.

In emergencies it may be necessary for the crew to jettison stores to reduce drag or hazards prior to an emergency landing. It may be possible to eject AAM prior to arming. Rail-launched AAM may have to be retained unless the launcher is on a jettisonable rack or pylon.

Environment

The environment to which a missile mounted on a high-performance aircraft is subjected can be very stressful for hardware and electronics. The AAM airframe must be designed to withstand these extremes with no loss of structural strength or system performance. Flight at exceptionally high altitude can expose the missile to temperatures as

Internal carriage of stores predated the development of stealth aircraft with weapons bays to reduce the radar signature of the jet. This flight test F-22A shows AIM-120 missiles in the port belly bay. Such designs do not simplify missile employment nor necessarily reduce the flight environment impacts on the AAM. (Author's collection)

Missiles that cannot be jettisoned may need to be retained in an emergency landing like the AIM-9 on the wingtip of this F-16A following a November 1997 landing during which one main landing gear collapsed. Mishaps during takeoff leave no opportunity to jettison stores. The possibility of the missile firing or exploding during such an event is very remote. (Israel MoD)

low as -75°F (-60°C). High supersonic flight can bring high airframe temperatures via friction with the passing air, leading edges of aero surfaces reaching as high as 1,600°F (900°C). Transonic airspeeds can mean shock waves 'dancing' about the exterior with alternating pressure waves across missiles. Even AAM mounted internally, in weapons bays, can suffer similarly with varying pressure gradients with doors open such that, where these are too extreme, deployable spoilers need to be installed at the front of the bay to moderate internal flow. The vibration induced by operation of aircraft onboard equipment – especially engines (changing with throttle setting) – or unsteady airflow will produce loads (stresses) on the airframe and potentially degrade internal elements. Even a round kept in a weapons bay will be subjected to vibrations and, with the bay doors open, potentially harsh airflow conditions. Engines and other devices like pumps produce noise as waves transmitted through the air that also impact the missile. Flow fields change based on flight condition and presence of adjacent stores during external carriage that can also change with download of stores. Sound pressure levels can be especially high during engine afterburner operation. During ground operations the sound waves will reflect off the surface and surroundings like buildings in potentially complex patterns. Sudden throttle retardation can cause a high-pressure slug of air to be 'belched' out of an engine inlet that then travels down the length of the aircraft, potentially impacting a missile. Separation of adjacent stores can produce shock loads as well.

The internal temperatures of the missile must also be addressed. Given that components have to operate while baking in a bright sun to more than 120°F (50°C) or much hotter from air friction in flight, or cold soaked at high altitude, the demands on the missile's internal environmental controls are not fixed. The vacuum tubes initially employed generated considerable heat that had to be dissipated via forced air (ram air from outside or fans) or circulated coolant like hydraulic fluid or a dedicated fluid. In other cases, electronics and other components must be heated to ensure suitable performance. Heaters consume considerable current, draining batteries. The moisture content of the air can, under certain conditions, produce condensation within the AAM that must be accounted for, especially as this can freeze and expand at high altitude. Contaminants in the air can degrade the performance of subsystems exposed to them, especially salt spray for those on aircraft aboard carriers or flying low above the sea.

Internal vibrations can be equally harmful and so must be anticipated such that design and assembly can reduce deleterious effects. Components can be mounted on rubber isolation dampers to ensure against resonance with the most dominant vibration modes present (see Chapter 2). Vibrations are particularly damaging to

Armeé de l'air Mirage F.1Cs stand alert at a French airbase armed with Super 530F radar-guided AAMs underwing and R.550 Magic infrared-guides missiles at the wingtips, plus cannon. The instances of actually getting airborne on a 'scramble' were few and with flights less than an hour, the Magic was designed for just 40 hours 'on-wing' time and the 530F similar. Training flights were conducted with inert rounds. In later years, with new generation fighters, missions could include long patrols with aerial refueling in which this short service life challenged the longevity of the rounds. (Armeé de l'air)

Especially early AAM had many fragile internal components, and even later weapons could suffer if abused. Handling of a missile in the maintenance shop, storage facility, transport to the aircraft, and loading all offer opportunities for harsh treatment that can degrade operability and service life. This image of an F-106A at Edwards Air Force Base shows an AIR-2A Genie air-to-air rocket on a trailer with large pneumatic tires and suspension system that should give a gentle ride over most airfield surfaces and hangar door rails. By contrast, the GAR-3 and GAR-4 Super Falcon missiles are on small dollies (that can be used to lift the missile up to the launch rail), sitting in hard mounts and rolling on small, hard rubber wheels that had to give a rough ride. (Albert Grandolini collection)

vacuum tubes and account for much of their high failure rate of early AAM. Protecting electronics from degradation via vibration and condensation can see liberal coatings of silicon potting compound or even encasing the circuit board stack in hardened resin. Such measures can eliminate the potential for repair, requiring remove-and-replace of the entire board or assembly. Abrupt jolts and the decaying vibrations they generate can be induced in the missile during ejection, motor ignition, and motor cutoff that must be accounted for in the design. Similarly, construction must be such as to permit operation from the moment of launch at up to 50g and maneuvering up to 40g (variable with altitude).

The vulnerability of the missile to electromagnetic (EM) interference with its functioning must be assessed and tested. This implies an adverse impact on electronic behavior from emissions coming from the aircraft, other stores, the surroundings, and even from within the AAM itself. Especially shipboard, with many close emitters, and other aircraft airborne in a formation, the potential for EM interference is all too real. Likewise, the AAM must not adversely impact the operation of aircraft system. Changes to any of these emitters after certification for service may require some of the analysis and testing to be repeated. Intentional jamming of the missile by a hostile source is addressed in Chapter 4.

Non-Fighter Applications

Vulnerable non-fighter aircraft can be fitted to operate light AAM for self-defense. These have typically been attack helicopters more likely to be engaged by fighters carrying late-generation missiles able to track and engage the low-flying, low IR signature targets. The AAM integrated on these non-fighter platforms have been infrared-homing types, typically adopted from light air-defense weapons requiring little from the aircraft to be operated. The potential for a successful engagement is increased with the later generation of such products with highly sensitive seekers for all-aspect capability tracking. Installation of the missile launch rail and aircraft internal wiring plus controls and displays are necessary and nontrivial. The uninhabited combat air vehicle (UCAV) concept, armed with a variety of weapons, emerged in the 2010s. While principally focused on air-to-ground missions, destroying air defenses in the first portion of an air campaign, the potential to carry AAM remained part of these concepts that were bound to bear fruit by mid-century.

A French Army Gazelle is seen being used as a testbed for firing a Mistral light anti-aircraft missile as an AAM. Many such experiments were conducted around the world and some approved installations adopted. However, the role of the attack or scout helicopter makes the added weight and complexity generally unwelcome – and unnecessary if air superiority is achieved. (Author's collection)

Other missions that sought a self-defense capability from AAM have been assault helicopters and patrols/surveillance aircraft as shown by a P-3 firing a Sidewinder missile, an image from a video frame grab. (US Navy)

The US Navy tested the X-47B to explore the potential for an uninhabited combat air vehicle (UCAV) that could execute a variety of air combat missions. The potential for employing AAM has been among the UCAV roles examined by many countries. There is likelihood such a system will be fielded by mid-century. (US Navy)

Several AAM have become the basis for air-to-ground and ground- or sea-based air-defense systems. The surface-to-air missile (SAM) application is challenging given that firing lacks the initial speed of a launch aircraft and is projected up against the pull of gravity, greatly reducing range. These subjects lay beyond the scope of this book.

Cost

Missiles are always going to cost many times more than a rocket and shell in development, production and sustainment. This is tolerated because of the promised greater assurance of destroying or driving off an opposition aircraft costing many times more than the missile, preventing that enemy warplane from causing considerable damage with its ordnance. However, this has also seen missiles costing half a million dollars or more shooting down drones costing a few tens of thousands of dollars. Additionally, as AAM effectiveness was shown to be much less than promised, with many AAM fired for each aircraft knocked down, the calculus has been cause for pause. Aircrew facing risk to their lives or having the potential for scoring a 'kill' would very rarely make a cost tradeoff analysis before firing missiles.

The insertion of ever-more advanced technology has continually driven up cost in parallel with effectiveness. Some 80 percent of unit cost goes into the target seeker and guidance sections of the round. That is also where most of the changes occur generation to generation, with the rest of the AAM potentially carried over from earlier models and types for a cost savings – either continued production or remanufactured rounds. However, as AAM approached $250,000 per round, lawmakers thought long and hard about these choices. Yet, they have consistently committed to new and more costly products over the 80-year history of AAM.

Obsolete AAM

All militaries desire to possess superior capabilities compared with potential adversaries – or, in the case of a superpower, everyone else in the world – as a measure of deterrence and, if war erupts, making the conflict short and with minimal loses via overwhelming or at least superior application of violence.[11] The worse circumstances would be that a mission need goes unfulfilled and the warfighters know they are at a disadvantage with respect to their likely enemies. This can arise through government negligence, failure to anticipate needs (usually via intelligence collection/analysis failures or disregarding the findings), programmed upgrades yet to be implemented, or lack of economic means. (Wealthy countries are more likely to

The AIM-7 Sparrow air-to-air missile served as the basis for the anti-radiation Shrike air-to-ground weapon, a transition accomplished for several AAM types. This was developed by the USA during the Vietnam War to hit SAM radars in air defense suppression strikes. The AGM-45 Shrike is seen here on the outboard wing station of an F-105. Inboard is an AGM-78 Standard anti-radiation missile created from a shipboard anti-aircraft missile. (USAF)

A large number of AAM systems have been revised to serve as SAMs for ground- or sea-launch, such as the Sea Sparrow at top (note reduced size aft fins and folding wings) and the Chaparral based on the Sidewinder missile at bottom. Both systems were progressively upgraded as new versions of their parent AAM emerged. (US DoD)

possess 'cutting-edge' weapon systems like advanced AAM, especially as these exceeded a quarter million dollars each.) Alternatively, the adversary can abruptly introduce a markedly superior and destabilizing system or countermeasure that dramatically shifts the combat 'equation.' The only compensation may be sacrificing aircraft and lives to achieve national objectives should war occur, or application of some counterbalancing system in other combat areas. Although no 'shooting war' has been launched because a superior or destabilizing AAM was about to be introduced, this is conceivable.

As AAM became the principal means of engaging aircraft air-to-air, and multiple generations of missiles were available or fielded simultaneously around the world by the 1980s, many services were keenly aware that they were at a severe disadvantage depending on who they might face in combat. Such circumstances mean that the nation could expect its very expensive air arms to be neutralized in a short time. Responding with effective and less costly countermeasures could balance the combat inequity. After 1990 and collapse of the Soviet Union more of these asymmetrical conflicts emerged where one adversary quickly asserted air dominance, or the lesser combatant yielded the sky while relying on ground-based air defenses like SAMs and anti-aircraft artillery (AAA) to defend ground assets.

National ambitions to develop and manufacture indigenous AAM can, depending upon resources applied, yield a system a generation or more removed from the best available elsewhere, produced by leading military arms-producing nations. This can happen because of political pressures surmounting military requirements. The air service could then be saddled for a decade or more with an inferior system depending on the combat circumstances they are drawn into.

Atrophying Capabilities

By the start of the twenty-first century, the Cold War well in the past, the rate of technological advance affecting AAM had slowed and the urgency of introducing superior missiles had retreated. New models and types appeared at a much-reduced rate. Development teams that had always been busy with 'the next best thing' became under-utilized and were reassigned to more productive tasks. New talent could not be grown as confidently as in the past. Resources for design and testing of AAM became seldom utilized and the cost of maintaining them could not be recovered by new programs. These might be 'mothballed' or neglected such that reconstituting them when needed becomes prohibitively costly. Additionally, the very high cost of developing an AAM means that these are unlikely to be undertaken without certain government funding authorized by the national legislature. Fielded systems might not be updated as advisable. While computer modeling and laboratory simulation can compensate to some extent, the lost capabilities and talent inevitably affect the ability and timeline of developing a new AAM when the time comes. Deficiencies requiring costly and time-consuming remedy prior to fielding become more likely.

Foreign Operators

With government authorization, defense manufacturers may sell AAM developed under national military funding. They may develop a weapon for a foreign country though, again, only upon government approval. Such restrictions are to contain capabilities of militaries in volatile areas of the world or to prevent critical defense technologies falling into the hands of potential adversaries via third-party sales, espionage, or combat circumstances permitting examination of the weapons. These 'leaks' or 'spills' can provide insight into foreign technological advances, leading to reverse-engineering and adoption of desirable features, or devising more effective countermeasures. To prevent this inadvertent technology transfer, the items approved for sale to some foreign operators (excluding close and reliable allies) might have reduced capabilities as an alternative model. Even disregarding this, the technology in a new weapon may be so sensitive as to require tighter security measures to be imposed to help prevent theft. These could be demanded as a term of sale, but with no certainty they will be followed rigidly. Technology acquisition via capture or theft of new weapons has been a real and persistent threat, with especially China accused of this practice as a means of rapidly 'catching up' with the West. This can subtract years and considerable expense in developing a new weapon while eliminating a hard-won advantage of the originating nation which must work all the harder to compensate.

Foreign operators often receive support from the originating nation and manufacturer as terms of sale or via ongoing contracts. The risk is that the originating source ceases to operate the system, cuts off the support for political reasons, or the manufacturer decides not to support it under business rationale. The operator may then face the potential of a weapon becoming inoperable over time. The only means of avoiding this is to establish local production of common replacement parts and to build up a stock of components to sustain the system throughout its anticipated service life. Buying up stocks from other nations is another method, although the seller may be bound by contract to seek the approval of the originating nation.

Attitudes

Some small nations are provided AAM by more affluent partners in a period of wealth and fervor that can fade, and the armed forces became lightly funded or demoted to low regard. Even first-rate forces can suffer periods of low performance owing principally to poor leadership. This can lead to degraded routines of care and maintenance of weapons, sometimes becoming the norm, undermining reliability of the systems. When officers do not appear to care, neither will the enlisted personnel who maintain such weapons. The missiles, like the aircraft, become poorly maintained such that the air force becomes more a symbol than a functional and reliable armed defense/offense capability ready to meet national exigencies. The AAM are slowly rendered unlikely to operate and are more for show as part of a 'make pretend' air force of 'boys with toys.' Combat results elsewhere may be disappointing, casting doubt on weapon effectiveness. Once aircrew lose confidence in their weapons the game just becomes a very bad joke.

A fine example of AAM export is Sweden's acquisition of the IR-homing AIM-4C (adopted as as the HM-58) and the radar homing AIM-26B (HM-55). These were also produced under license within the country by Saab-Scania as RB 28s and RB 27s, respectively. These weapons are shown beneath a Swedish Air Force J 35F Draken, the AIM-26Bs on the fuselage and AIM-4Cs under the wings. (Albert Grandolini collection)

Air-to-air missiles can be as challenging to maintain as the combat aircraft employing them. This image of a Mali MiG-21bis with R-3S and R-13M missiles stored in a dusty hangar, and a RS-2US missile laying on the ground, would cause many an armament specialist to shudder. So neglected, the rounds are likely unreliable, and the pilot can but give a prayer before firing and hope for divine intervention. (Tom Cooper collection)

2

MISSILE DEVELOPMENT

Regardless of country of origin or working agency/company, there is a common process for developing weapon systems such as AAM. Some elements of the process might be accelerated, run concurrently, or given short shrift, but these are fundamental. At best the process requires years to run and produce an operational capability. There may also be more than one program underway to meet the combat need, or multiple competitors for a single system, helping to ensure that if one falters there is still potential that the other succeeds. Programs may also be cancelled because funding falls short, the development runs too long, the system emerging appears to offer too little capability, the threat recedes, or the mission need changes.

The Process

The principal elements of an AAM program are these:

1. Combat capability deficiency identified – current system aging, threat capabilities not matched or exceeded, mission requirements altered, new mission arises
2. Requirements for new or modified weapon drafted
3. Assessment of available technology to meet requirement performed
4. Requirements broken down into measurable specifications ('specs') guiding design and test
5. Funding programmed and allocated
6. Fundamental technology maturation undertaken as needed
7. Contractors selected and contracts let
8. Preliminary design performed, reviewed and approved
9. Preliminary hardware and software assembled and tested
10. Requirements adjusted based on what is identified as possible, practical and affordable
11. Down-select to a single contractor, if relevant
12. Detailed design performed, reviewed and approved
13. Prototype elements assembled and tested; changes introduced as necessary
14. Prototype full-up round (FUR) assembled and subjected to systems ground testing
15. Low-rate initial production to provide test rounds and continue contractor funding, initial production line established
16. Fit checks of prototype FUR or mockup on handling equipment and aircraft
17. Static jettisons on the ground to check functionality and weapon dynamics
18. Captive carry flight testing conducted to assess ability to withstand environment and check interfaces with aircraft
19. Safe separation tests at various flight conditions and maneuvers
20. Boresight (unguided) shots off-aircraft to assess basic functions and safety
21. Guided shots against minimally maneuvering decoys or simulated targets
22. Guided shots against more challenging decoys (signature and maneuvering)
23. Mock combat scenario shots to test operational suitability, possibly including countermeasures
24. Production decision made and production line matured
25. Support systems and logistics 'tail' established, training performed, and any capital construction undertaken to field weapon
26. Full-rate production deliveries to combat units
27. Initial Operational Capability (IOC) achieved
28. Monitoring for system deficiencies in the field to assess need for alterations
29. Upgrades/maturation performed, with production line adjustments and follow-on testing, as needed

An army of engineers, technicians, machinists, and other personnel are required by the prime contractor and vendors to develop, manufacture and support AAM. In decades later than this image the desks would be dominated by computer monitors. (San Diego Air & Space Museum)

This list has evolved over time and, in the early years of AAM development, did not include all the steps or they were not distinct. Especially operationally representative testing was not usually very rigorous. Low-rate production decisions have been made before much testing was complete to ensure some cash flow to the developing company. These risk delivering rounds that do not fully meet requirements and need costly upgrades later (if possible) to bring them up to the operational standard.

Requirements are based on mission need – defeating potential adversary aircraft in the face of their defenses and countermeasures. However, the ability of the existing aircraft to carry the missile and integrating it with the aircraft systems may impose limitations such as weight, dimensions, mounting and separation method, or available supporting systems (electrical, electronic signals, cooling, and the like). These could, in turn, affect warhead mass, maneuvering capability and range. The expected timeline of development may also prompt some capabilities requiring more time to perfect to be set aside or postponed to a later model because they will delay fielding the AAM beyond the identified need date. The old dictum that the last 10 percent of performance requires 50 percent of the development time and money, especially if this represents the technology advancement, is simplistic but illustrates the tradeoff. Available funding may be another impediment to the inclusion of some desired features. Instead, these are introduced on the production line over time via 'block' upgrades or those already in the inventory via modifications by the manufacturer on recalled rounds or by field organizations (military depots or arsenals). This permits the weapon to be introduced at the earliest practical date to counter opposition capabilities and gain experience with the system while improvements are pursued. Software changes can alter jamming invulnerability, improve seeker performance like discriminating decoys, plus improve guidance and control, all without hardware changes, yet must be undertaken with care. Even introducing a 'firmware' change (circuit board) can be disruptive to the production line and remove rounds from service.

Program Termination

Weapon system development programs fail for many reasons. Among these are funds starvation, failure to meet requirements, poor contractor performance (commonly over-budget and behind schedule), failure of the technology (repeated test failures), and because the original requirements no longer meet warfighter need owing to the evolving 'battlefield.' All of these can be avoided to some extent with vigilant program monitoring plus clear and regular reporting to the legislature appropriating the funds. Risk reduction can include having different contractors working on different aspects of the missile matching their expertise (motor, flight control/guidance, seeker), and possibly multiple contractors for high-risk efforts such as new technology. This last helps to ensure that if one contractor fails another may be successful and the schedule still met.

Top: Some elements of missile production have required old-fashioned machining as showed here. By the 1960s numerically controlled machine tools were being used for greater precision. Later still it evolved to constructing lighter-weight composite materials components manufactured in more sterile environments. (San Diego Air & Space Museum)

Bottom: Much handwork has been involved in AAM production as demonstrated by these guidance sections from the 1960s being meticulously assembled and checked. Later computer-controlled machines would perform such tasks for greater precision and reliability, but some hand assembly remained necessary. A few hundred missiles per year have been a typical output. (San Diego Air & Space Museum)

The goal is an effective weapon produced profitably by a manufacturer and delivered on cost and on schedule to warfighters in required numbers. This last has evolved over the decades as weapon effectiveness has risen and the length of a military contest has seemed to draw down with the size of standing forces. Hughes Aircraft production of the Falcon and Super Falcon missile, in its many models, ran for nearly 20 years and tens of thousands of rounds – numbers that became rare by the second decade of the twenty-first century. This is suggested by the packaging area of the Tucson, Arizona, factory where missiles are being prepared for delivery and then an F-89H aircrew inspecting the AAM installed on their aircraft prior to a sortie. (Albert Grandolini collection)

The data from the terminated contract can be transferred to the remaining team provided the contract specifies such terms.

Compensation

Other aspects of the combat 'system' may compensate for a compromise in the new missile's capabilities. These can include radar detection of a target at longer range, deployment of new countermeasure systems, superior training via awareness of adversary weaknesses, and gaining the most from aircraft maneuvering/flight conditions to evade or gain a favorable firing position. (This last emphasizes the continued importance of these aspects even in the era of 'assured kill' AAM.) The mission need is still met in a timely manner via such trade-offs. Aircrew combat experience as compensation is an advantage that fades rapidly across a few years and can be relied upon only briefly as a benefit during extended periods of combat.

Service Life

A missile is designed for a predetermined lifetime based on anticipated need for a new AAM or simply what is achievable at a reasonable expense and timeline. The life is based on wear-and-tear of the equipment, stress of the flight environment ('on-wing' time), duration of exposure to the environment (flight profile and number of sorties anticipated), seals that degrade, the natural longevity of components like batteries that can be effectively recharged only so many times, and so on. The life can be extended by swapping in new or reconditioned components, provided the vendor can be induced to continue manufacturing them once stock is depleted.[1] Even if just stored for years and never mounted for flight on aircraft, there are typically periodic checks and replacement of 'shelf-life' limited parts to ensure readiness when needed.

A change to the aircraft and mission profiles can alter the anticipated service life and affect the effectiveness of the AAM. In some cases, the adverse effect is only evident in combat – the worst possible outcome to be avoided at all costs. These changes can include extended periods in low-altitude/high-speed flight where more buffeting is encountered, more high-g maneuvering, operation of new onboard equipment or carriage of new external stores/configurations that can alter vibration and shock levels impinging on the AAM, introduction of new adjacent stores that alter the flow field around the missile affecting separation trajectory, and the like. Operation in a different EM environment can degrade electronic performance, produce self-jamming or unintentional decoying.

Testing Regimen

Testing the AAM, from the smallest component to the FUR, is vital to ensuring the product meets specifications so that the contractor can be paid in full and the warfighter gets the expected capability. Some components are delivered from vendors to contractor specifications and expected to have been tested separately. This may include the seemingly minor parts like screws to more substantial elements like integrated circuit 'chips.' Yet, all have the potential for undermining performance and reliability of the FUR if falling short of specs, and so the contractor may perform their own tests.

Generally, assemblies are tested in stages up to the FUR. These trials include laboratory ('lab') tests, ground tests and flight tests. Some ground tests are precursors to flight trials or are dedicated events deriving data not achievable in flight. As a principle, testing proceeds from least stressful to most stressful for the system in a buildup fashion such that risks of irrecoverable failures are reduced to avoid cost and schedule impacts. Depending on the government's 'mood' and public visibility of the program, testing failures can become highlighted as emblematic of a 'troubled' program and risk termination. This can stress all involved and produce undesirable consequences. Failures are, in fact, almost certain to occur and need not be catastrophic if hazards are mitigated. Yet, each failure provides data essential for perfecting the system. With testing

This flight test F-16C is carrying an AIM-120 under the port wing along with an enormous air-to-ground weapon. Such an underwing assortment of stores affects the airflow field around the AAM that can alter separation behavior and so requires flight testing to assess safety and effectiveness. Heavy stores on the F-16 have historically induced wing oscillations that can accelerate wear of AAM components. The aircraft also has a parachute canister installed above the engine nozzle to be used if the aircraft departs controlled flight during high angle-of-attack (AoA) maneuvering and fails to recover. High AoA testing in a variety of configurations, including with AAM, have usually been found essential. (Author's collection)

requiring years of meticulous and costly effort, there is always the urge to find 'shortcuts.'

Many tests are undertaken by the contractor prior to delivery of rounds to the government. The government procuring agency, without a fiduciary interest in the outcome, is obligated, possibly by law, to undertake trials of the missile. Their results are generally considered the 'last word' in spec compliance to assess contactor performance (so that they can be paid in full) and AAM readiness for service. They also bring resources that contractors typically find unaffordable to acquire and sustain given infrequency of use. Especially vast open-air test ranges on or over public land are the domain of the government. Military personnel and property such as warplanes are also committed by the government. Many nations have dedicated test centers with such resources where contractors and the military can conduct the trials.

The trials may use unique rounds that substitute instrumentation in place of the warhead and fuze to measure various quantities via transducers, possibly telemetry (TM) to transmit the measurements to a ground station, and recording systems if the round is retained or recovered. Initially the engineers were fortunate to get a dozen channels of TM'ed data. After the digital 'revolution' this increased continuously until thousands of parameters could be transmitted and recorded onboard.

With deficiencies identified at any stage of the testing regimen, corrective changes are likely to be introduced. This may require regression testing to ensure the alterations did not unexpectedly affect other aspects of the system already evaluated. Such is commonly performed initially in the lab and ground tests before return to flight trials as risk-reduction progression. It is also critical to maintain configuration control such that multiple changes introduced nearly simultaneously are well documented and

A modest-size flight test mission control room is typical of this National Aeronautics and Space Administration (NASA) facility from the 1980s, used for an X-29 aircraft sortie. The data monitoring stations evolved from pen-plotters to electronic displays over the decades, and the volume of data from a handful of parameters telemetered to the ground to thousands, but the role and function of a control room has remained unchanged. (NASA)

understood so that when new issues arise the source can be isolated via fault tracking and isolation without confusion. All of this is constrained by time, schedule, available resources and capabilities, operational environment realism achievable, and funding.

Laboratory Testing

All components of the emerging AAM are bench-tested and perfected under controlled conditions in the laboratory. Results characterize performance and seek to identify deficiencies to be remedied at the earliest opportunity. This includes components integrated in the lab and tested with simulated flight conditions and target behavior for optimization and troubleshooting. Considerable laboratory equipment and computing resources are essential for such work. Software is also tested on computers prior to integration with firmware as elements of test articles.

The buildup can proceed from open-loop to closed-loop tests. Open-loop implies an input injected into an element or subsystem and the output measured to match against expectations. Closed-loop means that the output is fed-back, per design, into elements to alter output to be assessed against expectations. The buildup may begin with simulated elements, using unique computer programs written by contractor programmers, to actual hardware-in-the-loop. The buildup includes ever-more complex scenarios such that the system is characterized end-to-end, and confidence gained that the FUR will meet specs.

Vital lab tests are conducted to verify compliance with service life requirements. The flight environment impinging on the AAM (Chapter 1) can be modeled mathematically, if suitably characterized, to guide the design. Flight and ground testing may be necessary to collect data characterizing this environment and then guide designers in ensuring the product will survive for the required period of service. This means an instrumentation and recorder package installed where and how the eventual missile will be transported and carried, and data gathered during representative aircraft operation. The environment is reproduced in the laboratory to qualify prototype component, or the FUR, to verify the AAM can withstand the conditions. Apart from characterizing performance and uncovering deficiencies, results are also used to inform mathematical models of the system for analysis at condition that cannot be practically simulated and at many more conditions than cost and schedule permits.

Component testing within a laboratory setting is conducted to verify spec compliance, optimize performance, and resolve deficiencies. This Sidewinder seeker (likely from the AIM-9D) is in such a setting, the component held in a very stiff fixture to ensure alignment to simulated targets within fine tolerances. The attached gas hose no doubt delivers coolant and the gas cylinder the motive pressure for the moving parts. The individual is Dr. Bill McLean, onetime lead for the Sidewinder program. (Albert Grandolini collection)

Component operation in exceptionally high and low temperatures and pressures are tested in laboratory chambers. The missile or components are also mounted on a 'shaker table' where they are subjected to various frequencies and amplitudes of vibrations in all three axes of motion over long periods. Everything resonates at certain natural frequencies and, if subjected to such a frequency, the amplitude of structural response (deformation) increases markedly. This can lead to structural fatigue – promoting material cracking that undermines load-bearing capability – or component failure over time (breaking, no load-bearing capability). A coupling of two resonant modes at like frequencies can produce an unstable structural oscillation that can grow to catastrophic amplitudes. Coupling implies a similar frequency and a phase relationship that enforces the resonant oscillations. This is called flutter and design analyses then construction seeks to avoid the instability, which is then checked in flight test. The acoustics environment can be simulated in resonant or reverberation chamber tests with the missile or components subjected to sound overpressures of various frequencies, amplitudes and directions. Several lifetimes of exposure to these environmental factors can be simulated in an accelerated fashion using such tests. Specifications may demand more than one lifetime be demonstrated to cover uncertainties in predicting the environment, inability to simulate same with suitable fidelity, and how representative the missile or components are to the final in-service round. Periodic tear-down inspections and component functional checks during the testing seek to identify failures not readily evident.

Ground Testing

A rocket motor produces significant sound pressure levels and heat at ignition and launch. This must be measured, and results checked against the design specifications of the nearby aircraft structure, systems and stores to ensure against detrimental effect even for the split second the AAM is on or near the aircraft at firing. Reverberation off the aircraft back onto the missile must also be considered. Similarly, the sound levels generated by the aircraft engine and the launch of other ordnance can adversely affect the AAM during carriage and so must also be measured for comparison with the design specs. This includes ground operations with sound reflected by surrounding structures.

Scaled models or the full-size external shapes of AAMs are often placed in a wind tunnel to collect aerodynamic data on lift and drag plus the motions these induce on the body that the flight control system must be capable of reacting. The effectiveness of the flight control surfaces is also measured. By the twenty-first century it was possible to derive much of this analytically and greatly reduce the amount of costly wind tunnel time required, or eliminate it entirely. A sled track has also served to get the missile to speeds unavailable in a wind tunnel.

The aircraft integration ground tests will include fit checks of at least a dummy round to verify suitability with ground transport and loading equipment, aircraft launchers, ground crew tools, adequacy of written instruction, etcetera. The crew will also note any difficulty they encounter in handling the missile. Changes will be made to optimize this aspect of the integration. Systems support from aircraft to missile, data transfers, and correct status indications will

Some equipment used to perform environmental qualification testing of missile components are seen in this laboratory, euphemistically called 'shake and bake.' At right is a 'shaker table', at left possible an acoustics chamber, and left foreground possibly a pressure and temperature test cell. (San Diego Air & Space Museum)

Wind tunnel scale models of an aerodynamic shape such as a missile must be precisely fabricated for the most accurate data to be derived from the test. Shown is a metal model of the Navy's Diamondback developmental missiles, an image likely from the late 1950s with the machinist checking dimensions with a micrometer. (US Navy)

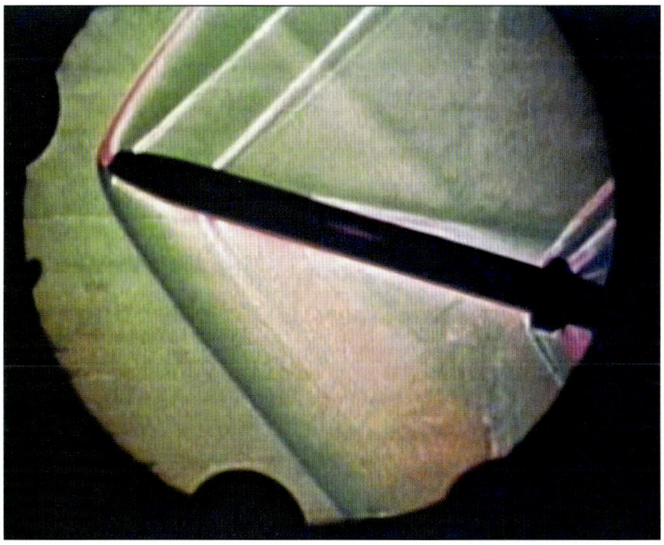

Most AAMs fly at supersonic airspeeds which generate shock waves across the exterior. These can have profound effects on stability and control and so must be modeled. Wind tunnel data can be vital in this task. This image shows the model of a US Navy AIM-95 Agile missile with shock waves visible via a special optical means (Schlieren). There is also the suggestion of low pressure atop the shape for 'body lift' to help in executing sharp turns for high agility. (US Navy)

Placing the full-scale or scaled model of the missile in a wind tunnel permits aerodynamic and motion data to be collected that then guides design. This image shows the Ryan XAAM-A-1 Firebird missile undergoing such tests, possibly at full scale. The mount can reorient the missile and also measure forces and rotations. (San Diego Air & Space Museum, Ryan Aeronautical Collection)

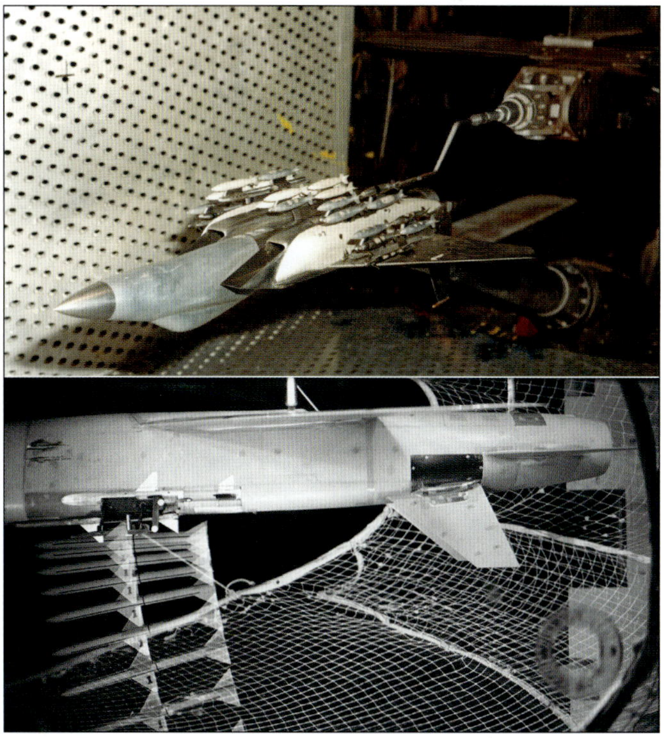

Determining the influence of the aircraft and other stores on the aircraft during AAM separation is done analytically, through ground test, and then in flight trials. These wind tunnel tests show, at top, a modeled store being moved away from its mount on a scaled F-15E as if in separation. The wind tunnel 'sting' moving the shape measures the airflow's influence on the weapon from which motion can be determined. At bottom a free-drop model of a Sparrow separating from a partial model of an F8U-3 and falling into a net. The time-lapse photograph shows the separation trajectory to assess safety. (Top author's collection, bottom Jay Miller collection)

Computational fluid dynamics permit aerodynamic parameters to be determined analytically with the aid of computers and mathematical models of airflow for a subject submerged in that flow. This model of an RF-4E shows the pressure distribution across the surface of the aircraft and stores at a specific flight condition via the various colors. The results were only as good as the models ('garbage in, garbage out') but came to be relied upon more and more after 2000. (Author's collection)

be checked, possibly with failure modes injected. Static ejections (if designed for ejected launch) will be into a pit below the weapon, the cavity lined with styrofoam or rudder material, sometimes with boxes that will be crushed, to protect the AAM against damage. This is called a 'pit drop' or 'box drop.' The round is instrumented to record imparted forces and motion while the event is filmed at high speed to assess the motion. These data are feed into analytical math models of release dynamics that support predictions for in-flight releases.

The radio frequency (RF) emissions from missiles and their reaction to surrounding or directed RF energy (EM interference, Chapter 1) must be characterized. This provides data to ensure compatibility with other aircraft systems and invulnerability to unintentional or intentional jamming. This characterization is performed as a ground test in an anechoic chamber. The chamber is an RF 'quiet' environment with material to absorb stray emissions. The emissions from the missiles, including frequency, strength, and direction, are measured by sensors within the chamber. For example, radar sidelobes (Figure 10) can be mapped. Emissions of various characteristics are then directed at the missile and reactions recorded. These data may lead to design adjustments or changes in employment tactics. Similarly, target reflections can be simulated to assess tracking capability. Jamming signals are generated to adjust electronic counter-countermeasure (ECCM) algorithms. Additionally, boresight errors owing to refraction of the RF signal through the radome material (aberrations) can be measured for mathematical correction within the guidance system to reduce this contribution to miss distance.

Test articles on rocket-propelled sleds running on miles-long tracks are another means of collecting aerodynamic and systems data at representative speeds for missile development. Hughes sought data on their AAM using the sled track at Edwards Air Force Base in the California high desert (10,000ft track shown at bottom). At top the full-size basic missile shape for the GAR-3 is prepared for a run on 20 January 1960. Inset is a developmental model of the Sidewinder missile propelling itself down the track at Naval Air Station (NAS) China Lake. (USAF Air Force Test Center and US Navy)

A bomb and AIM-120C missile are prepared for static ejection from an F-35B weapons bay into a pit. The pit may have boxes or foam to cushion the fall. Such tests exercise the ejection system and demonstrate safe separation from aircraft components and other stores, at least at zero airspeed and no aircraft maneuvering. This is done prior to in-flight safe separation trials. Data are also collected to support predictions of behavior at flight conditions. (US Navy)

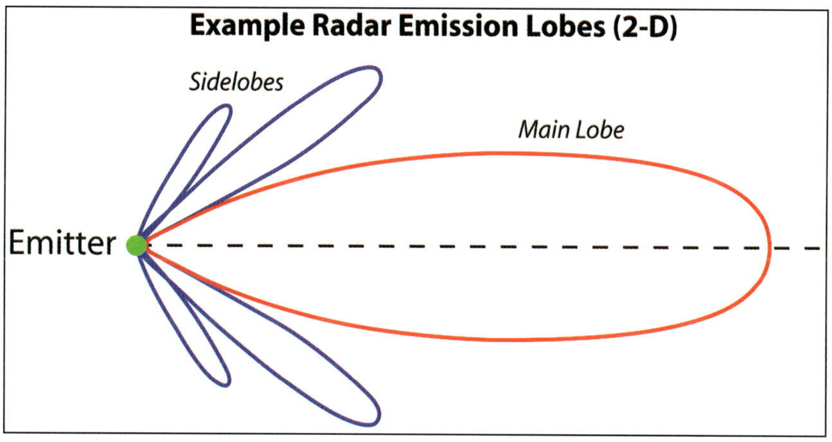

Figure 10: Radar emission main lobe and sidelobes are shown in this two-dimensional drawing. In actuality, it would have a three-dimensional character. These are characteristics of aircraft radar, missile radar, and RF fuzes. (Drawing by author)

A missile seeker head is mounted in an electronically 'quiet' anechoic chamber ahead of tests to measure emissions and effects of impinging signals. The foam pyramids lining the walls of the chamber and the mounting hardware reduce reflection of EM waves. The seeker head mount is articulated to change orientation with respect to transmit and receive antennas within the chamber. (San Diego Air & Space Museum)

Open-air radar seeker tests collect data on its reaction to angular motion glint reflections off a maneuvering target (Figure 11). This employs a Radar Target Scatter (RATSCAT) facility with the seeker mounted miles or fractions of a mile from the reflective subject, usually an aircraft or model thereof, sitting atop a mount constructed of non-reflective materials. The mount, built well above the ground, is on a rotating platform to change the orientation of the subject with respect to the seeker. Another open-air ground-based test has targets passing in front of the seeker, or any type, mounted in a facility or a roof of a high building and so exercises acquisition and tracking functions. The targets can pass at different ranges and angles while also exercising countermeasures.

Table 2 shows the many characteristics of the RF return from a target, some of which contribute to miss distance for radar-guided AAM as well as effective guidance.[2] It is, therefore, prudent to quantify these characteristics and compensate where necessary in the design to meet operational requirements.

Table 2: Characteristics of Target RF Returns

Characteristic	Definition
Reflectivity	Effective radar cross-section as seen by seeker and contributing to power of return radar signal
Doppler Frequency	Change in RF frequency owing to closing velocity
Glint	Lateral displacement of return owing to reflection hot spots and aspect angle
Amplitude Scintillation	Variation in power of return owing to reflection hot spots and aspect angle
Polarization	Reorientation of the signal waveform owing to reflection
Ground Clutter	Reflections off the ground from main beam and/or antenna sidelobes
Jet Engine Modulation	Doppler frequency returns from rotating engine components inconsistent with aircraft main return

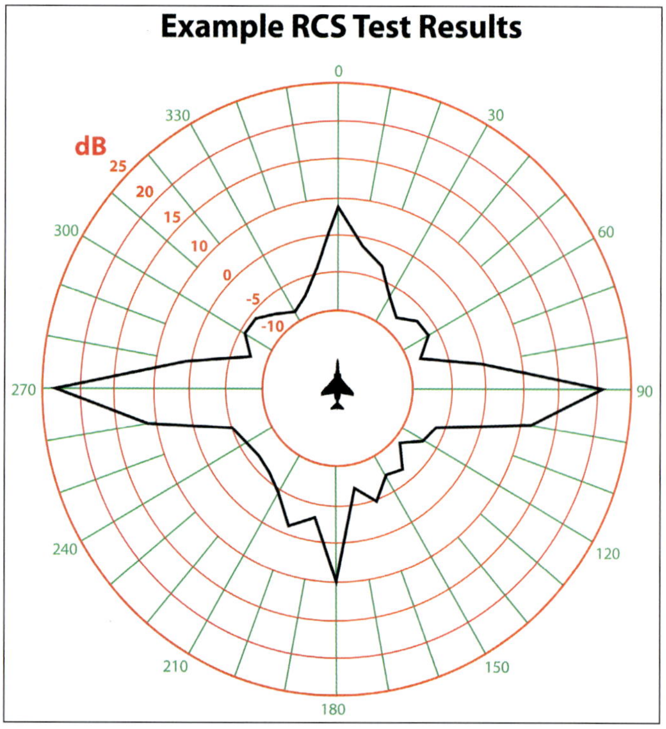

Figure 11: The results of a radar cross-section (RCS) test is shown for a horizontal aspect around a full 360deg of a specific aircraft for a specific radar type and power. The scale is decibels (dB), in this case running from -10 to 25dB. The amplitudes would vary with changes in aircraft configuration (external stores dropped, speedbrakes deployed, etcetera), different aspects in the full sphere around the aircraft, and different modes and power of the radar. (Drawing by author)

Flight Testing

A fundamental limitation of the flight testing of a missile, free from the launcher, is that it is a brief one-time event that typically ends with destruction or loss of the test article. This makes for a costly event for which as much data must be obtained with as high a certainty as possible. Such tests, then, are comparatively few and so backed by extensive laboratory and ground trials.

Free-flight data without the errors induced by wind tunnels can be had from ground launch. This could be powered initially off a rail or launch tower via the integral motor or atop a booster rocket. The data would be telemetered to a ground station for recording and analysis, or recorded onboard and downloaded following the flight provided the missile has a recovery system (typically parachute and perhaps floatation bags if firing over water). In the early days these

An open-air Radar Target Scatter (RATSCAT) facility measures an AAM seeker's reaction to radar reflection glint. The aircraft or model (F-102 shown) is mounted above the ground on an assembly that would normally be covered in radar non-reflective material and, as can be seen, the entire arrangement is on a rotating circular platform. The radar seeker under test may be miles away in a building where its response is monitored and recorded. (US Air Force)

onboard recordings could be pen-plots on paper or scratched on a revolving cylinder, all needing to be recovered for manual analysis. Another method was to photograph onboard instruments (called a photo panel) and the film recovered post-flight for tabulation of the observed readings versus time. Hand analysis moved to mechanical computers and then digital computers and calculators with ever-more capacity and speed.

Initial flight testing can be performed with the missile boosted off the ground atop a rocket and then then released for free flight. While this technique fell from use by the 1970s, direct firing of the missile off a ground launcher continues to be a useful method to test motor, stability and control. Guided shots have also been performed by guiding to a target flying above the test site.

Components of the missile, especially the inertial navigation system (INS) and the seeker head,[3] may be installed on another platform for flight testing in a more representative environment than a laboratory. This may be on a surrogate missile or a testbed aircraft where component responses can be measured and recorded. Particularly radar seekers may undergo flight testing against a variety of targets with the unit mounted fixed on a flight platform. Free-flights of radar-guided AAM can be conducted with false targets inserted into the guidance system via test-unique equipment. However, the surrogate aircraft provides the opportunity for many simulated intercept scenarios to be flown for a comparable wealth of data.

The development flight testing proceeds via buildup with the intent to clear the operating envelope and exercise all aspects of the system in characterizing performance, verifying specification compliance and uncovering lingering deficiencies to be corrected. Tests of the basic carriage effects on the aircraft and missile also offer the opportunity to checks the interface between the aircraft and the missile. Safe separation tests may be ejection of an inert weapon or firing unguided (boresight shot) off a rail. Testing the separation is a vital employment of the integration and ensuring safety against adverse dynamics endangering the aircraft. Missiles that are to be separated in flight are usually given unique markings to permit their motion to be determined via cameras mounted on the test aircraft or a 'chase' aircraft, or more rarely a ground tracking station. Apart from assessing if the missile is rotating, the application of mathematics (photogrammetry) can precisely model missile trajectory and distance from the launch aircraft. Flight trajectory once away may be recorded by a ground camera mounting on a cinetheodolite mount with manual tracking by an operator.

India employed ground launch of their Astra ARH missile during a long development period to affordably collect flight data. One such event is shown, likely at the Chandipur range, near the Bay of Bengal, during the 2000s. This test technique has been used from the earliest days of AAM development. (DRDO)

Missiles have been recovered following their tests by such means as parachutes. This permits examination and possibly download of data recorded onboard. This image is of a Firebird missile (booster rocket separated) following a test at Hollman Air Force Base (AFB), New Mexico, during 1948. The parachute did not prevent damage upon landing, possibly due to excessive surface winds. Wind limits are usually imposed on such tests. (National Archives via Tony Landis)

A JB-3 Tiamat missile is launched to flight speed atop a booster rocket in 1947 at the USA's National Advisory Committee or Aeronautics (NACA) Wallops Island facility. The technique allowed data to be collected in flight without the complexity and hazards of firing from an aircraft, especially early in development. The image shows the four swept fins adopted at one point in the testing of this experimental missile. The rocket appears to be tailing a wire back to the zero-length launch platform. (NASA)

Above: Capturing and relaying telemetry signals from the missiles and support aircraft, and tracking these via radar, over dozens or hundreds of miles can demand substantial facilities. Shown is a US Navy installation on Tranquillion Peak, Point Arguello, California in 1961. The data may then need to be relayed via radio or landlines. (US Navy)

Left: The IR seeker head of a developmental missile, or with new elements derived from R&D, is attached under the nose of the USAF's NC-131H (53-7793) testbed as photographed in the mid-1980s. Such surrogate testing permits the seeker to be exposed to the actual in-flight environment and targets arranged during flight testing for data collection. This is more controlled and effective than firing a missile against a single target with telemetered data. The shown fixed installation probably supports repeated fly-by of ground IR source targets. (Author)

Above: An AIM-9X is shown on a wingtip station of the F-35A during clearance testing. The target-like marking on the weapons, adjacent stores and aircraft are optical references to be used in photogrammetry determination of motion during launch, extracted from high-speed camera recordings. (US Air Force Test Center)

Left: Test ranges have often employed cameras tracking the test aircraft and missile manually (cinetheodolite) and via radar. Much data can be derived from the resulting film. In some cases good telemetry lock is also dependent on such close tracking. (NASA)

Guided shots may, for radar-guided AAM, initially use targets simulated within the seeker or against radar-reflective articles aloft on a towline or on a balloon. Infrared-guided AAM can be fired against parachute flares. Testing then proceeds to firing against inexpensive drones before proceeding to more complex drones, usually converted outdated fighters, maneuvering and employing countermeasures. The difficulty and complexity of the test scenarios grows to the ultimate trials of firing multiple rounds against multiple targets, with countermeasures employed, in as realistic a combat environment as can be safely replicated.

In some cases, the flight test round may be a singular and precious resource worth more than the carrying aircraft. Some test articles have instrumentation and modified flight systems supporting specific test objectives. The test pilots may be strongly discouraged from jettisoning the store in an emergency. The testing also requires a large test range given the potential for the missile to go awry and the hazards this could represent when it reaches the ground. This may require the manufacturer and funding agency to test far from home. The missiles are commonly modified with a radio-commanded destruct feature substituted for the warhead such that a range safety officer can command a flight termination (sends the signal for the package to explode) should the round appear to be going astray.

The flight testing can be hazardous when performing functions for the first time given the potential for unexpected and unsafe results. This is especially true when testing failure modes cases or extreme conditions. Considerable safety measures are taken, and the test monitored from the ground and a safety chase aircraft nearby. One Hughes Aircraft manager, participating in planning a particularly hazardous Phoenix missile test with the F-14 Tomcat aircraft, recalled the lead-up and consequences:

> The Navy expected Grumman, as manufacturer of the Tomcat, to verify that missiles could be safely launched from the aircraft while it was maneuvering in a reverse to normal – that is, while “pushing over negative g's,” an unusual maneuver, rarely used except in emergencies. In this test, the plan was to be at minus 3 g (negative gravitational force). Whether or not the ejection launchers would have enough force to physically separate the weapon before rocket motor ignition was uncertain; in this condition, a weapon carried beneath the airframe would have a force three times its weight trying to push it back against the F-14's fuselage.
>
> Grumman and Raytheon agreed to try with the Sparrow missile, but our analysts calculated that the Phoenix missile would not safely separate in such a maneuver. I had to make the unwelcome telephone call to the Deputy Chief of Naval Operations for Air (Navy 05), hopefully to prevent a test with our missile. Two weeks later, Grumman proceeded with the Sparrow. The missile's rocket fired while the weapon was not yet separated and its wing fin sliced the Tomcat's underside, cutting fuel lines. A giant plume of flames 300 feet long gushed aft; the pilot tried to save the aircraft, but that was not possible. Fortunately, both crew members ejected safely and were rescued from the sea. For a while, I had been in the customer's doghouse, portrayed as a “real wimp,” but after the Sparrow disaster our dissenting posture was regarded as a heroic stance.[4]

Some tests such as this may be judged too hazardous and the potential for occurring in operation too remote to assume the test risks. Such ‘corner point’ conditions are cleared by analysis and extrapolation from less hazardous or costly trial results. Cautions or prohibitions (placards) may be placed in the flight manual, though inhibiting the combat pilot and restricting the operating envelope is strongly discouraged. Over the decades modeling and simulation progressively reduced the number of costly captive carriage and shots required. However, this is reliant upon empirical data to ensure suitably representative mathematical models, either from prior programs or the applicable program. Some flight testing will always be necessary.

The final production configuration is a carefully balanced design. Any change to the missile, the platform, or the launch environment may create unexpected consequences undermining reliability and effectiveness. Regression testing is advisable to verify no adverse effects or to characterize performance in the new configuration. The extent of the regression work is dependent on the nature of the changes and the data collected during the course of the work. Consequently, testing can continue throughout the life of the system.

Service Evaluation

Close on the heels of development testing is the operational evaluation to assess the true combat capability of the weapon against the requirements and current tactics. The results of such evaluations, often with some character of tests, may be mandated prior to a full-rate production decision and the ‘big bucks’ the contractor has been waiting for. This may be undertaken with the understanding the missile has lingering immaturities and so the conclusions may be to assess whether it is ‘good enough.’ If it is no better than that which it is to replace or complement, it may as well not be produced. (National political pressures and internal service factors may still trump the evaluators' recommendations. People sometimes lose focus on the real purpose of the effort which is to improve combat potential.) Even beyond this ‘op eval’ are workups to determine the best tactics to exploit the weapon to best effect, especially against specific targets. The results, typically classified, are then communicated to operational units and training organizations. All such evaluations should be conducted without contractor involvement and strict checks that the missiles supplied are representative of the preproduction or low-rate production rounds and not special ‘gold plated’ articles via special contractor attention to ‘tip the scale’ towards higher probability of good performance.

Planning and executing the operationally representative shots can be as complex and expensive as persecuting an actual aerial engagement, especially for many-on-many scenarios (multiple shooters and targets). Such tests are vital in determining the true potential of the weapon, versus ‘brochure’ numbers, and insight into how best to fight with the AAM.

Reliability

Elements of the AAM will eventually fail due to poor workmanship, inadequate quality control, operation beyond the design specs, or just wear-and-tear over in-service time. Development component testing will seek to demonstrate reliability as specified in the requirements, usually defined as mean time between failures (MTBF) by methods discussed elsewhere in this text. The MTBF rates for components will be tracked during flight testing, ‘op eval’ and initial service. If the AAM is falling short of the specified reliability then remedial efforts may be taken. The data are also used to plan the number of components to purchase and stock for future replacement by field units or depots. The tracking of MTBF will likely continue throughout the life of the weapon to observe aging degradation for adjustment to spares procurement or possibly redesign and new manufacture of the least reliable components.

During initial service the weapons are repeatedly inspected, and systems tests performed to track natural degradation and identify emerging deficiencies or more rapid degradation than anticipated. The system is intended to last a planned service period, generally decades, and anything undermining this must be addressed promptly. Aging systems are also subjected to periodic inspections and tests to ensure against unanticipated degradation or to evaluate system upgrades seeking to prolong combat effectiveness.

Support Equipment and Logistics

All unique equipment and tools required to support the system in service must be included in the AAM acquisition. This may be as simple as a hand tool and as complex as an electronic console to check the AAM or its components in a maintenance shop. The logistics 'tail' for the weapons, its parts, and the support equipment plus documentation must also be set up to include programming delivery of all such elements across the service life of the system. The operator needs to integrate all this into existing organizational structures.

Training

Those supporting and operating the AAM require familiarization and then specific training. Training aids need to be designed and assembled for this purpose. In some cases, these are the actual support equipment with features to simulate actual in-service procedures and equipment responses. Captive carriage training rounds are also produced to more safely conduct flight training. This usually substitutes a recording device for the fuze and warhead to support post-flight debriefing of exercises. These may also lack the fins and other aero surfaces for simplicity of loading. The motor may be replaced with ballast. Training rounds typically have a unique finish or a colored band indicating their nature. Inert dummy rounds are usually painted blue. Other colored bands may indicate specific models of an AAM type, or the nature of the warhead fitted.

Technology Insertion

Legacy weapons can be improved as underlying technologies evolve. The simplest examples of this is electronics miniaturization and microprocessor capabilities, improvements in batteries, and the evolution of radar technology. The American Sidewinder missile is evidence of this with evolution through many models across half a century seeing transition from vacuum tubes (thermionic valves) to transistorized circuits (so-called solid-state), the introduction of integrated circuits with semiconductor-based microprocessors, and then computerized technology with software with that improved via high-speed processors offering ever-more capability in smaller and smaller volume. Such changes can sometimes mean significant electrical system changes; especially as vacuum tube circuits demand high voltage with low current whereas solid-state circuits are low voltage/high current systems. Software can also impose processing latency of hundreds of milliseconds of delay compared with essentially instantaneous analog systems.

Considerable savings in budget and time can be realized if these improvements are adopted as 'drop-in' swaps of existing elements with new to create a more capable AAM model without a wholesale redesign and production line rework. In the late twentieth century the 'spiral development' approach became more accepted by which fundamental and affordable capabilities are

Aircraft that operated with the SARH Falcon AAM, in its many variants, could carry the Weapon System Evaluator Missile (WSEM). This mimicked the AAM functions prior to launch and recorded the events. As seen conformally mounted on the F-101B rotary weapon door and held by servicemen of an F-102 squadron, it had only strakes instead of wings. (Jim Hawkins collection via Dennis Jenkins and San Diego Air & Space Museum, Jack Machun collection)

sought for initial fielding and immediate military advantage while follow-on pre-programmed development permitted introduction of progressive improvements over many more years. The weapons must be designed in a modular configuration to permit components to be swapped as more advanced versions became available for refinement and progressively greater capabilities. In some cases, new software 'loads' can introduce the enhancement. This, however, is not effortless or fast as software testing and some ground trials are required, at a minimum, these consuming months and considerable resources to ensure against unexpected consequences while verifying the expected results. The notion that 'software don't break' is misleading.

Foreign Suppliers

By the late twentieth century nearly every country required raw material from abroad to fabricate complex and advanced weapons like guided missiles. Additionally, some piece-parts of an AAM, including microprocessors, come from foreign sources. All of these are subject to embargo at the whim of the foreign supplier government or as part of an alliance, interference with their transport by another nation via blockade or sanctions, or suspension of production for simple commercial rationale. Even superpowers are subject to such factors that could affect development and production of AAM until alternatives can be identified and substituted – many months or years. Consequently, efforts to ensure all is locally produced, like the USA's 'Buy American' policy, are largely political showpieces.

Disposal

Safe and environmentally friendly means of disposing of weapons became a significant factor in development during the late twentieth century. This must be considered from the first and can affect design and material choices. Toxic materials, especially, must be identified, proper handling documented, and disposal means selected.

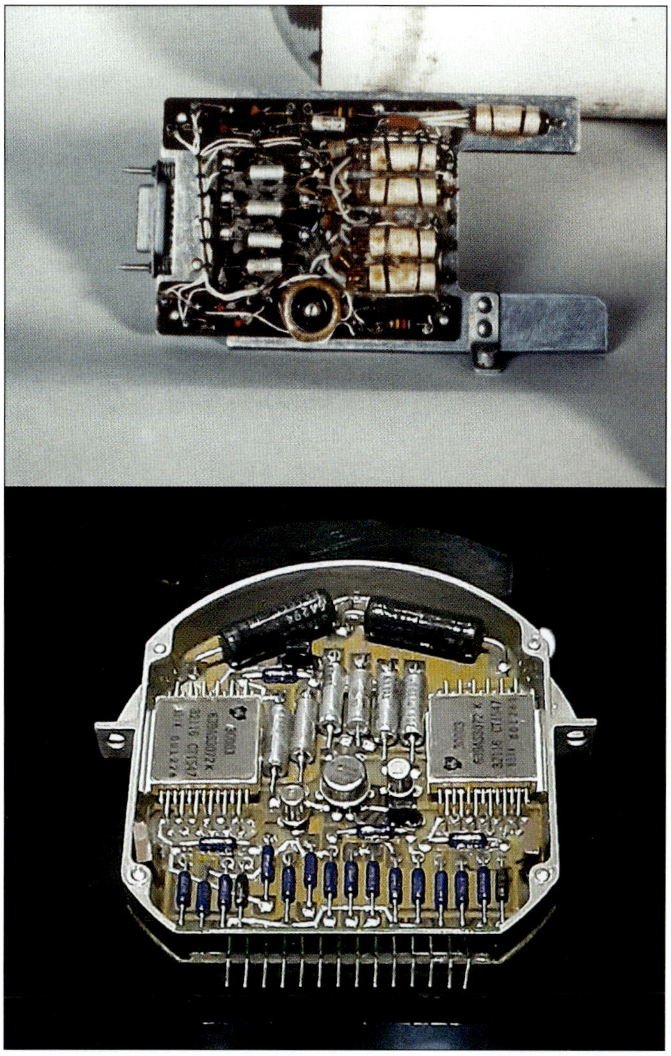

At top a circuit board from a Soviet R-3R missile shows that these were essentially hand-built in the mid-1960s. The USA had done the same initially, but by that time had begun to automate many processes. At bottom is a Sidewinder component from about the late 1970s assembled in that manner. (Top US DoD, bottom author)

The evolution of electronics across the history of AAM played a vital role in improving capabilities and reliability. From vacuum tubes to high-speed microprocessors requiring much to little cooling, the vital 'innards' of the missiles shrank and became more ruggedized while offering more potential. The guidance section elements of a displayed AIM-4D missile reveal the mid-1960s technology with stacked circuit boards. These have a great many metal-encased vacuum tubes, capacitors and resistors all hand soldered to printed circuit boards and assembled in a light chassis for dense packaging. (Author)

3

MISSILE ELEMENTS

Air-to-air missiles are commonly built in a modular fashion (Figure 1). While the seeker is at the front and the motor at the rear, the order of the other assemblies can vary. The various components may also be manufactured by different firms and designed to fit and operate together. All have some similar features, but great variation in details.

Aircraft Interface

As introduced in Chapter 1, the interface between the aircraft and the AAM may include passing power, cooling and electrical signals, usually via a cable called the umbilical. An example of such signals is presented in Table 3, some content of which will be understood only later in this text. Historically each signal has been passed via a dedicated wire suggesting, for those presented in the table, a minimum 11 wires in the cable and 11 pins in the connecting cannon plug. The digital 'revolution' permitted considerable missile self-diagnostics as built-in test (BIT) with reporting of this status via the cable. This may be as simple as 'ready' or 'fail', or more complex signals in a coded digital signal. At the missile end of the cable the connection is designed to be severed at launch motion (frangible) without disturbing the flight. Table 4 shows example data transferred from the aircraft's FCS to the AAM prior to launch of a SARH missile. The FCS uses the aircraft systems, such as radar and navigation, to assist the AAM in detecting and tracking the target prior to launch.[1] The 'initial conditions' in the table are for a late-model Sparrow III, but still 1960s technology. All are subject to errors given the limits of the technology and so affecting system accuracy. Table 5 shows the greatly reduced signals passed to an ARH missile from the FCS with 1980s technology and an inertial reference (introduced in Chapter 2). Both assume an aircraft equipped with a Doppler radar (see later).

The umbilical or umbilical attach points for a Sparrow missile (left), Sidewinder (center) and Falcon (right) are displayed. The Sidewinder is an AIM-9M training round and shows the umbilical in the main image along with its attach point that allows it to shear away on launch. Inset is a view of the face of the umbilical's cannon plug mating to the 'male' end on the aircraft's launcher side. The larger pin hole with the metallic sleeve is where coolant is passed to the AAM's seeker. The Sparrow umbilical pin has two such coolant ports, likely for delivery and one for return. Hughes preferred press contacts in an umbilical pad on the side of their Falcons, simplifying separation. (Jim Rotramel collection, author, and author's collection)

At right a sailor replaces the nitrogen canister in a Sidewinder launcher. This would cool the seeker prior to launch until exhausted, improving sensitivity. At left the sailor is attaching the umbilical to the mating female cannon plug mounted in the launcher, accessed by moving the forward cap of the launcher aside. (US Navy)

Table 3: Umbilical Signals[2]

No.	Signal	Function
1	Missile Ground	Missile grounding and power return
2	Launch Command	Denotes missile launch status
3	Battery Activate	Inhibits missile battery operation
4	Fuze Out	Inhibits warhead fuzing when the missile is on the launcher
5	Electrical Power	Provides power for missile heaters and operation prior to launch
6	Missile Identification	Indicates to aircraft stores management computer presence of the missile. Used in determining hung status, missile priority, and missile power application functions
7	Battery Armed	Missile provides battery armed indication when this reaches its specified voltage
8	Motor Fire Command	Provides command to missile to ignite motor
9	BIT Initiate	Initiates missile BIT
10	Missile Status	Missile status as determined by BIT and provided to stores management computer
11	Slave/Cueing Commands	Cues or slaves seeker to an off-boresight target or out of FOR

Table 4: Sparrow III Missile Initial Conditions[3]

Condition	Signal	Function
1	In-Flight Switch	Switches autopilot gains during flight as function of target/interceptor altitude differential
2	Dogfight Command	Sets servo loop gains to accommodate short range targets
3	Sweep Control	Sets velocity gate to proper orientation after launch
4	Roll Command	Sets missile roll to proper orientation after launch
5	Sweep Select	Selects sweep mode for acquisition
6	English Bias	Provides yaw course correction prior to acquisition
7	Roll Gain Switch	Adjusts roll autopilot to high gain as function of launch station
8	Pulse-Doppler Command	Indicates aircraft radar mode
9	Head Aim Pitch	Indicates pitch direction to target for aiming missile
10	True Airspeed	Aircraft velocity used in determining missile autopilot settings
11	Head Aim Yaw	Indicates yaw direction to target for aiming forward antenna
12	Range at Launch	Range to target for predicting terminal gain switching and fuze arming
13	English Bias Pitch	Provides pitch correction prior to intercept
14	Altitude Switch	Adjusts autopilot gains as a function of launch altitude
15	Simulated Doppler	Aircraft-measured closing velocity Doppler signal for missile sweep positioning

Table 5: Active Radar Homing Missile Initial Conditions[4]

Condition	Signal
1	Missile Mode
2	Active Frequency Designation
3	Target Position
4	Target Velocity
5	Target Doppler
6	Aircraft Orientation Axis Angles
7	Aircraft Velocity
8	Aircraft Position
9	Aircraft Altitude

Tables 3, 4 and 5 indicate a process that must be completed before the missile leaves the aircraft. The general steps and timeline are illustrated in Figure 12.[5]

Table 3 refers to slave and cueing and is illustrated in Figure 13. The high off-boresight (HOBS) capabilities of later generation missiles made it useful to cue the seeker head via other onboard sensors (or the aircrew eyeballs) prior to firing, especially when portions of the aircraft blocked the seeker FOV and prevented normal acquisition. This could be the radar, infrared search and track (IRST) system, television display, an operator slewing a designator on a wide-angle head-up display (HUD, likely via a control on the throttle), or a pilot helmet-mounted display (HMD, also helmet-mounted sight) with the target designator moved by the turning of the pilot's head and then selected by a hand or voice control. Alternatively, the pilot's LOS via the HMD can slave the radar which, in turn, cues the AAM seeker. The head movement is precisely tracked by a system within the cockpit, and this is converted into bearing and elevation data. This 'look and shoot' capability ensures that the AAM has acquired the intended target before launch (lock-on before launch, LOBL) or at least initially steered immediately after launch such that it will acquire the intended target after launch (lock-on after launch, LOAL). Such systems must be kept within calibration tolerances for them to function effectively in cueing a seeker with narrow FOV.

By the 1980s it became necessary to standardize aircraft data protocols (formatting and organization) and the MIL-STD 1553 (Military Standard) was devised in the USA for databus designs. A databus is a computer function for distributing data (digitally coded electronic signals) to the various elements in the aircraft to include downlink to an AAM. The missile, then, also had to conform to the protocols or have some conversion function. The MIL-STD 1553 protocols were adopted by most manufactures in the West, especially as they sought to market weapons for integration on American aircraft. There are similar protocols in other manufacturing nations and alliances.

Airframe

The AAM has a basic structure to support its internal and external components and to react loads imparted during launch and flight. The airframe must be strong enough to withstand ground handling, takeoff and landing of the aircraft (including shipboard catapult and arrestment) separation forces, thrust application, maneuvering loads (load factor and aerodynamic pressure distribution), any impulses imposed from gusts encountered, altering pressure distribution due to shock waves, and high temperatures from the friction of supersonic flight through the air. The most obvious aspect of the airframe is the outer surface (moldline) that carries aerodynamic forces (pressure distribution, called airloads). There may be internal structure just as important such as stiffeners for the outer skin, ribs and

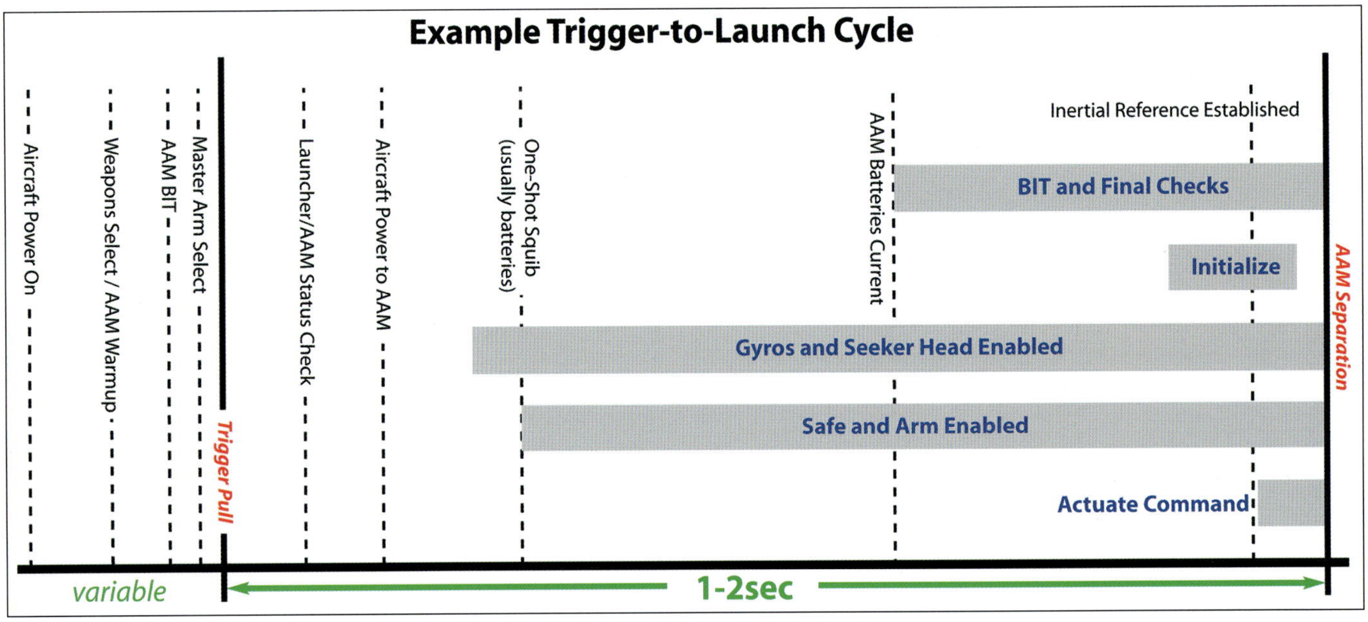

Figure 12: An example trigger-to-launch cycle shows the common actions and automatic functions that the aircrew and systems go through before the missile fires. The 1–2sec duration shrank over the decades, dependent on the particular AAM type, but the general concept remained the same. (Drawing by author)

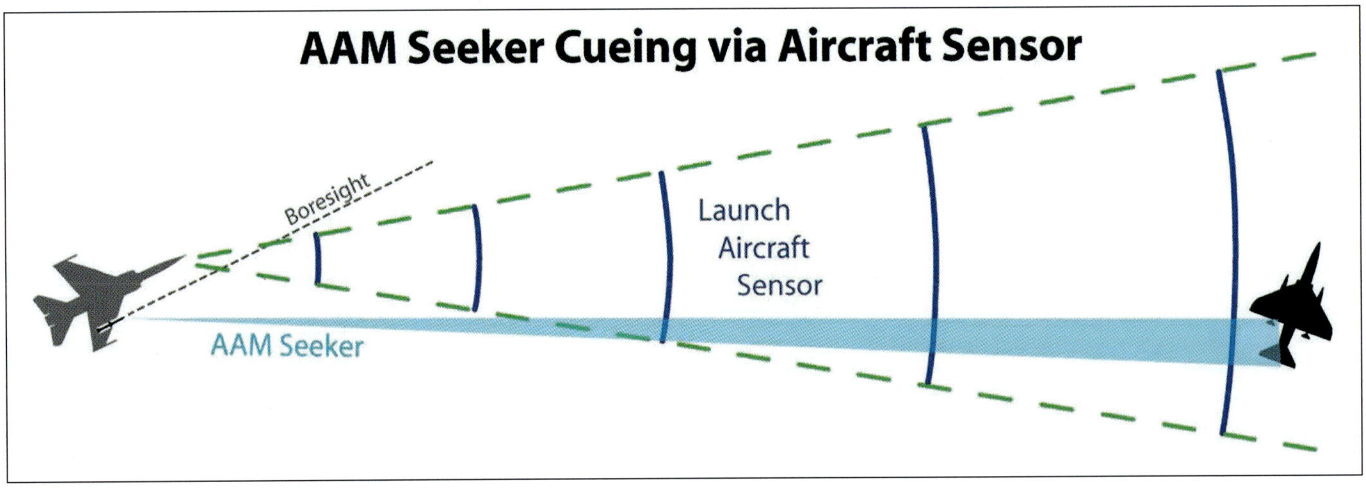

Figure 13: This is an illustration of an aircraft sensor detecting the target and directing the AAM seeker is directed well off-boresight (cued or slaved) and acquires the target prior to launch. The target is initially detected on an aircraft system like the radar, pilot helmet-mounted display (HMD), or infrared search and track (IRST) systems. (Drawing by author)

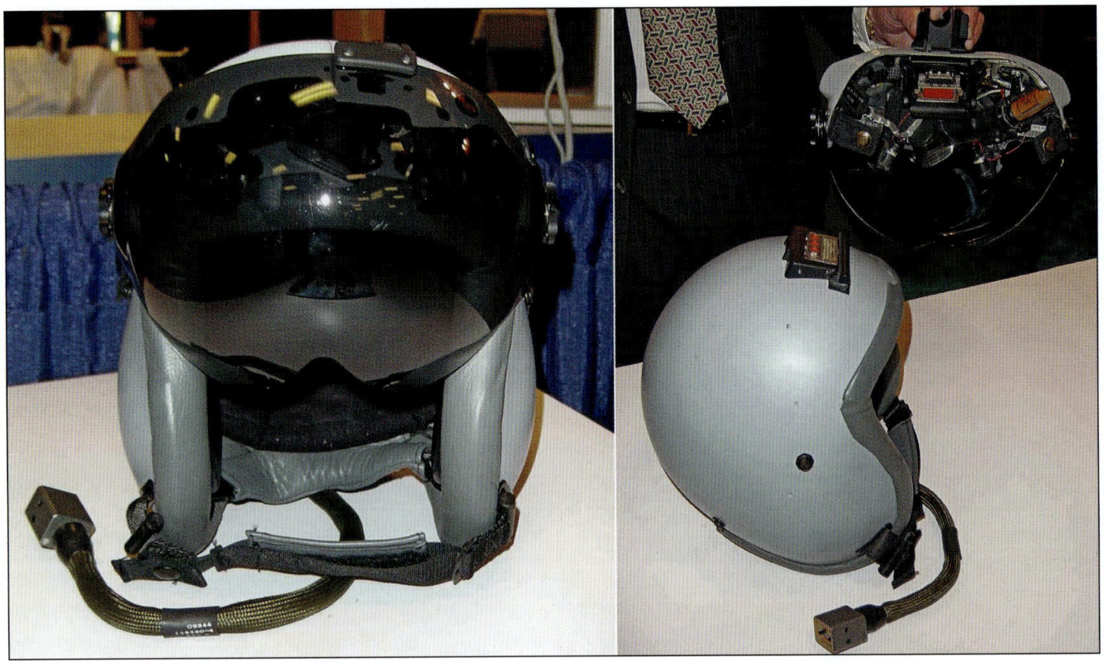

The Joint Helmet-Mounted Cueing System (JHMCS) is an example of HMD equipment that allows a pilot to direct missile seekers to a target it might not otherwise detect, such as to its off-boresight limits. The helmets are usually characterized by a large package of gear at the front top, requiring care in reducing weight and ensuring balance for pilot neck comfort. The inside of this package is shown for the JHMCS, including means of projecting images on the visor. Such systems require modifications to the aircraft to which they are intended to interface. (Jim Rotramel)

longerons, engine mounts, and the like, but such semi-monocoque construction became rare in AAM during the 1950s. Most missiles use the outer skin (casing) alone as the main load-bearing structure (monocoque construction). This is commonly a metal or composite fiber tube. Airframe flexure under loads can create false motion sensed by autopilot sensors and inertial references, apart from altering the external aero configuration ever so slightly, and so must be considered if not minimized to the extent possible. The output of those devices, if periodic, can be electronically filtered to attenuate the adverse frequencies.

Being able to track and engage a maneuvering aircraft or missile requires superior maneuverability. This can be equated simply as load factor or g. The missile typically must be capable of maneuvering to a minimum three times the load factor limit of the target. This is accomplished by the autopilot with the flight control surfaces and sustained by a suitably designed airframe. The load factor sustainable is a function of flight condition – altitude and airspeed (Figure 14). The generation of fighter aircraft extant when AAMs matured maneuvered up to 7.2g. Later generations maneuvered to 9g or more. This suggests minimum AAM maneuvers to 35–40g, but much more in the latest generation.[6]

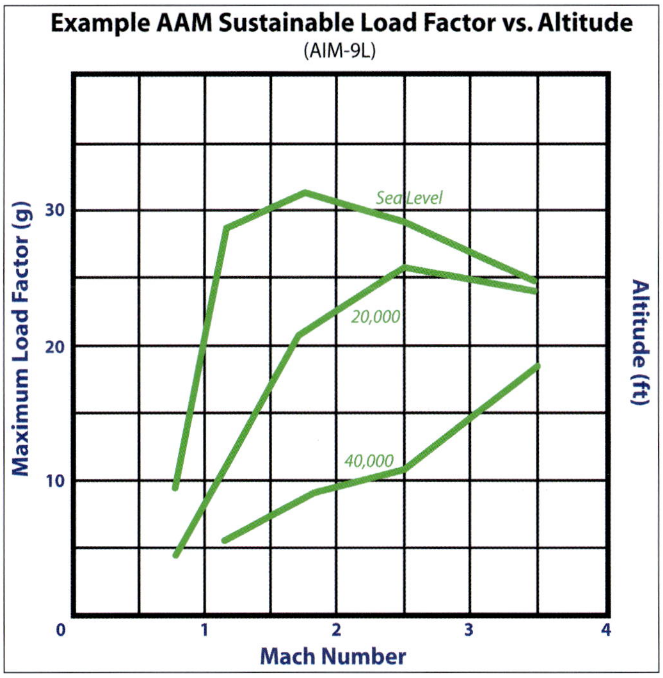

Figure 14: The sustainable load factors for the AIM-9L missile at three altitudes is shown with respect to Mach number. Clearly, the advertised maximum 30g is only sustainable at one condition, and an unlikely one at that. The general trend is greater sustainable load-factor with increasing airspeed and decreasing altitude. The shape of the curves is probably a function of shocks moving across the body as flight conditions change. (Drawing by author)

The basic cylindrical missile body with cruciform aero surfaces is generally favorable for low drag and ability to construct a stout structure. It is desirable to have the aft end of this tube close down to the outer diameter of the rocket motor nozzle to reduce boattail drag. The nose of this body is tapered, commonly ogive, to also reduce drag in supersonic flight. The IRH seekers commonly use a hemispherical window nose and so the desire is to make this the lowest practical diameter for drag reduction. Where the nose cone serves as a radome for a radar receiver or transceiver antenna, this must be formed of material reasonably permeable to the frequencies intended. Some energy loss and scattering of the energy, called aberration, is unavoidable but reduced to the greatest extent practical by material selection and thickness distribution, or compensated for computationally. It is frequently necessary to run electrical signals, power, and perhaps gases between segments of the missile, but penetrating intervening sections such as warhead and motor is impractical. In these cases an inter-sectional fairing (cable run, tunnel) attached to the outer case (Figure 1) encloses these elements to prevent snagging and to reduce drag.

Motor

The function of the rocket motor is to propel the missile off the aircraft and towards the target. Thrust is measured as specific impulse – thrust per unit mass of propellant. This is increased by different compositions of propellant chemistry. The motor, as a whole, must withstand very low temperatures at altitude and flight vibrations (also during transport and storage). The combustion chamber must withstand the intense heat and pressure of motor operation. All such are considered in motor selection and other design choices, some which will have a cascading effect on other aspects of the missile, such as weight and cost. There is no 'free lunch' in aerospace design. For any propellant type, propellant consumption (burning) alters the weight and mass distribution of the missile, and the flight controls must be able to accommodate this.

The initial thrust is high (boost) to ensure a quick and safe separation from the aircraft to include accelerating the AAM to flight speed down a launch rail. In free flight the desire is to reach speed permitting safe and effective maneuvering as soon as after clearing the aircraft as possible. The missile must be clear of the aircraft because it may abruptly maneuver across its flightpath during engagement of the target. Subsequent propellant burning may be at lower impulse (sustainer) to maintain the flight speed as the missile guides to the target. At propellant exhaustion the missile continues flying on course while decelerating until reaching the target or controlled flight can no longer be maintained.

Missile nozzles are typically fixed rather than with vectoring or with an alterable exit area as used in space launch vehicles. It must not degrade under the intense heat and pressure of the combustion for the brief one-time flight. The classical convergent/divergent nozzle is designed to produce an exit pressure closely matching the design point flight condition – for example, standard static air pressure at 35,000ft/10,700m – to optimize thrust. Operation off the design point is common and will unavoidably generate less than the ideal thrust, though this is seldom significant. The nozzle throat dimensions must also seek to optimize the pressure within the combustion chamber for the most efficient propellant burn. The neck-down to the nozzle throat ahead of the expanding bell-mouth exit, also called a blast tube, offers volume within the AAM case frequently used for control surface actuators. Many designs lengthened the throat for this purpose, necessarily resulting in a loss of exhaust pressure by a few percent and so lower thrust.

The choice of propellant is generally based on intended mass and volume of the missile. Liquid propellants require tanks for the fuel and oxidizer to sustain combustion, pumps and fuel lines. This also offers the potential for throttling, though this is seldom essential for an AAM. Solid propellants are typically lighter, consume less volume (fuel and oxidizer together), and frequently deliver more thrust per unit weight (thrust:weight ratio). They cannot be throttled or shutdown and restarted. As with any motor, increasing the quantity of propellant to extend range means an increase in mass and volume and so demands more thrust to sustain performance in a 'death spiral' of engineering trade-offs.

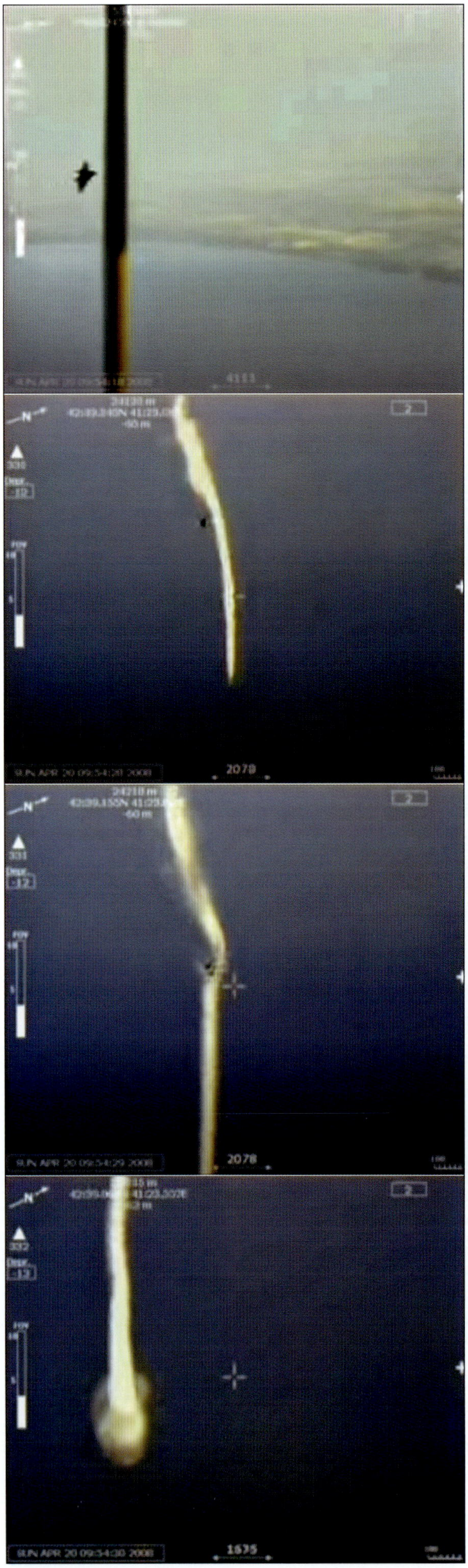

Two images of cutaway solid propellant rocket motors show features of the convergent-divergent nozzle, the case, and propellant grain. The top image shows a nozzle that is a separate assembly from the case but neatly fitting therein. The bottom example has an insert of a different material for the throat to reduce erosion altering dimensions during the burn. (Author)

The motor nozzle may have a plug to prevent foreign matter from entering during storage and ground handling. This is blown out on ignition. The AIM-9 would usually have this, initially a brass disk soldered into the nozzle throat. The rollerons at the aft fin tips (Volume 2, Chapter 1) have also had several means of 'caging' them from flopping about prior to launch that complicates transport and loading – in this case cords that are burned through at motor ignition. (Author)

Left: These series of images are frame grabs from the surveillance camera video of a Georgian drone as it was looked behind and recorded itself being shot down on 20 April 2008 by a Russian MiG-29 with an R-73 missile. The first frame at the top shows the fighter rolling in behind the drone at less than half a mile trail, likely right at the minimum firing distance of the AAM. (The object bisecting the image is a portion of the drone.) The second frame down shows the first fraction of a second of the missile's flight, leaving a long smoky trail behind the fighter (although it is also overtaking the trail) and as it races ahead. The third frame shows the MiG passing through and disrupting the smoke trail. The bottom frame shows the missile abruptly adjusting upwards for the final fraction of a second before detonating. (Public Domain)

By the third generation of missiles, efforts were made to reduce the smoke generated during combustion. 'Smokeless' solid propellants were typically achieved by reducing or eliminating metals from the grain. However, condensation from moisture in the air may still form a contrail and make the flight path visible as long as the motor burns. A smoke trail, if spotted in time, warns an adversary that they are under fire. After motor burnout there may still be a faint smoke trail from residual combustion. During the 1960s and 1970s the Soviets included flare 'tracers' at the end of many of their AAM so that the aircrew could see the fly-out after motor burnout, day or night. This lent some confidence the missile was performing as expected or, if not, was cue to fire another.

Solid Propellant

A solid propellant motor is the most common for AAM. It simplifies storage and no propellant loading is necessary in the field. The propellant, or grain, is usually poured or cast in a molten state into a metal or wound composite casing and adheres to the case (case bonding). The case then serves as the combustion chamber. The case may be inserted into the missile airframe or, more commonly, serve as the airframe section outer surface. Burning of the solidified propellant produces high-pressure gas that passes out the nozzle at the end of the case. The nozzle is typically mechanically fastened to the seamless case to accommodate the propellant pour. The propellant may be cast as a uniform column from top to bottom of the case or with a central open core (perforated propellant).

The motor casing must withstand the high temperatures and pressures of the burn. Weak areas and cracks in the case, or voids in a composite layup, can lead to pressure venting and motor explosion. The casings are generally examined via X-ray or other non-destructive inspection methods, like ultrasound, to ensure integrity before propellant pouring.

Loading of the AAM onto the aircraft requires an electrical connection through which the electrical charge used to ignite the motor propellant is carried. An electrical squib igniter can be at the end of the propellant, just ahead of the nozzle, for the uniformly cast grain. This base charge produces an end or 'cigarette' burn like a candle, from above the nozzle to the top of the case. More common is the hollow column up the center of the propellant cylinder for core burning to offer more surface area ignited initially. This delivers higher initial specific impulse. It also permits the igniter to be placed at the forward end (top) of the propellant column, usually generating a brief but intense flame down the length of the core to ignite the whole exposed surface area immediately. The propellant then burns outward. Unlike end burning, from end-to-end, this outward burning insulates the case from the heat until the final moments of combustion.

The inner surface of the propellant in a 'core burner' can be cast with a cross-sectional pattern, (Figure 15) to tailor the thrust profile. A simple tubular column would give relatively uniform thrust throughout the burn owing to the uniform surface area. A star pattern shown yields a non-uniform thrust profile, owing to varying surface area as the burn progresses. This typically provides an initial high impulse (most surface area being burned) to launch the missile and then may sustain the burn (more uniform area) at the peak airspeed achieved or slows the speed decay until the endgame when the AAM can execute sharp maneuvers with less airframe stress. As described previously, the initial portion of the burn is referred to as the boost phase or pulse followed by the sustainer. Another design is to have a grain poured and allowed to solidify before pouring a second and different grain atop the first (Figure 15). Different composition or density of grain produces different specific impulse and burn duration in addition to varying thrust from the surface area being burned. With end burning, the first portion of the propellant serves as the boost pulse (rapid combustion, high impulse) and burns to the point of igniting the second pulse as the sustainer (longer burning and possibly lower impulse). The two grains are burned in tandem with no pause between. An alternative is to have the first grain poured along the length of the cases and solidifying with a large central core in a more complex process. The second grain is then poured similarly but adhering to the first and leaving the common perforated core. This permits top ignition, the booster grain burning outward to ignite the sustainer. Yet another design has both grains burning simultaneously (Figure 16), but the booster grain is consumed fastest (possibly due to the core design) leaving the sustainer (possibly with only a simple column core) to continue burning until exhaustion.

These two defense workers hold a cast solid propellant rocket motor in a wound carbon fiber case. The rounded end (top of the motor) is common for pressure vessels. The nozzle is away from the camera. The weight of the component is evident. (San Diego Air & Space Museum)

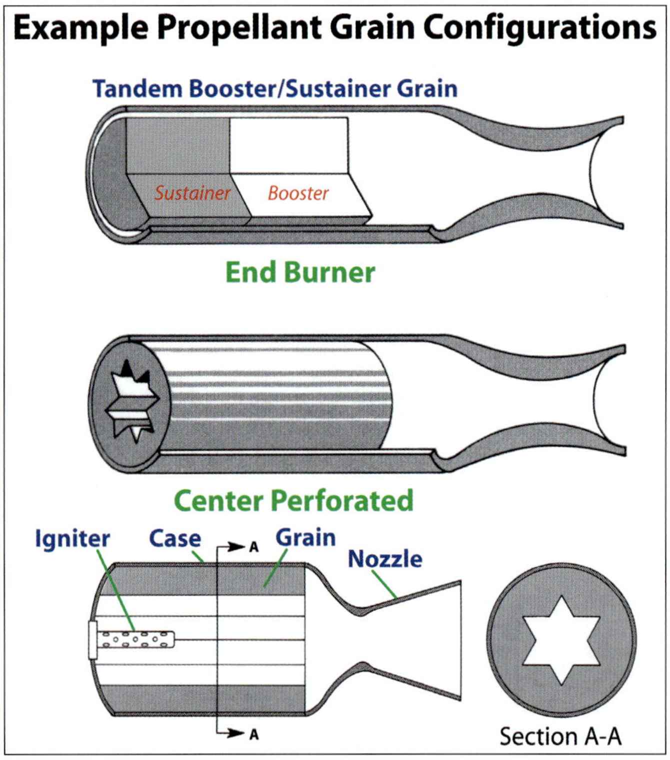

Figure 15: Cutaway drawings of a solid propellant motor shows, at bottom, a propellant grain with a perforated center (star pattern) and igniter, offering more propellant surface burning initially than the end burner option (top). The top drawing also depicts the example of a tandem booster/sustainer motor. (USAF, modified by author)

Many SAM rockets and launch vehicles employ stages of propulsion sections, the boosters dropping away when exhausted to reduce vehicle weight while the next stage begins operating. This has been very rare in AAM. An analogous design is a propulsion section with two motors in one case, in tandem, physically segmented by a barrier that disintegrates on the forward (second) motor ignition. The first pulse, just forward of the nozzle, is ignited with a base charge and the second pulse by a top charge (Figure 17). This arrangement allows the second charge to be ignited at the optimal point in the flight profile, as determined by algorithm or timer, while the AAM coasts until that time. A great increase in range and addition of energy for endgame maneuvering is the benefit. When first introduced, these motors were just mentioned as dual-pulse or two-stage, intentionally concealing their nature. In later cases the references were simply nonspecific.

All of these different grain structures affect AAM airspeed and duration of the fly-out, disregarding speed decay from maneuvering. These differences are depicted in Figure 18. A combination of segmented pulse motors or ramjets (see later) and new, more energetic propellant grains have made very long-range air-to-air missiles (VLRAAM, greater than 100mi/160km) possible by the 2010s.

Even the solid propellant motor needs to be inspected periodically and some rounds statically fired occasionally on a ground test stand to ensure the material is not degrading. Temperature extremes cause the grain to expand and contract at different rates than the case, potentially separating from the chamber walls and cracking.

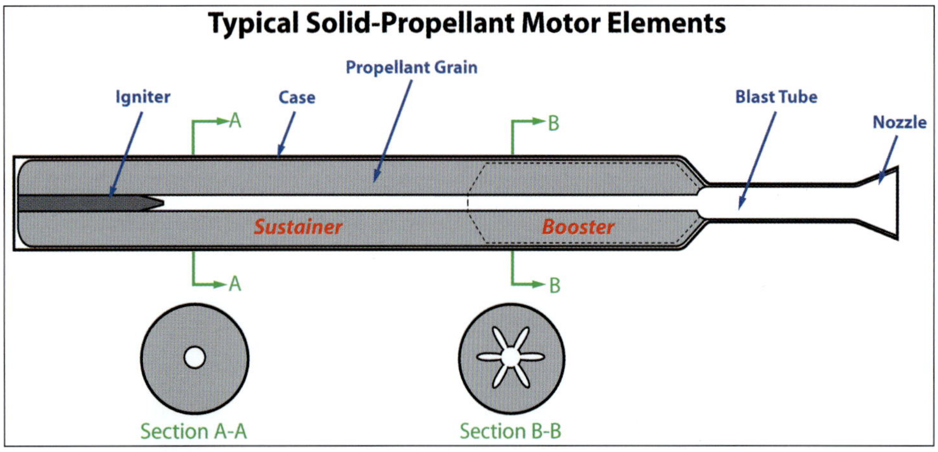

Figure 16: This solid propellant configuration has the booster and sustainer pulses as the same grain but only the center core design distinguishing them. Both are ignited simultaneously but the booster burns rapidly owing to its core configuration and shorter length to leave the sustainer to continue until burnout. (Drawing by author)

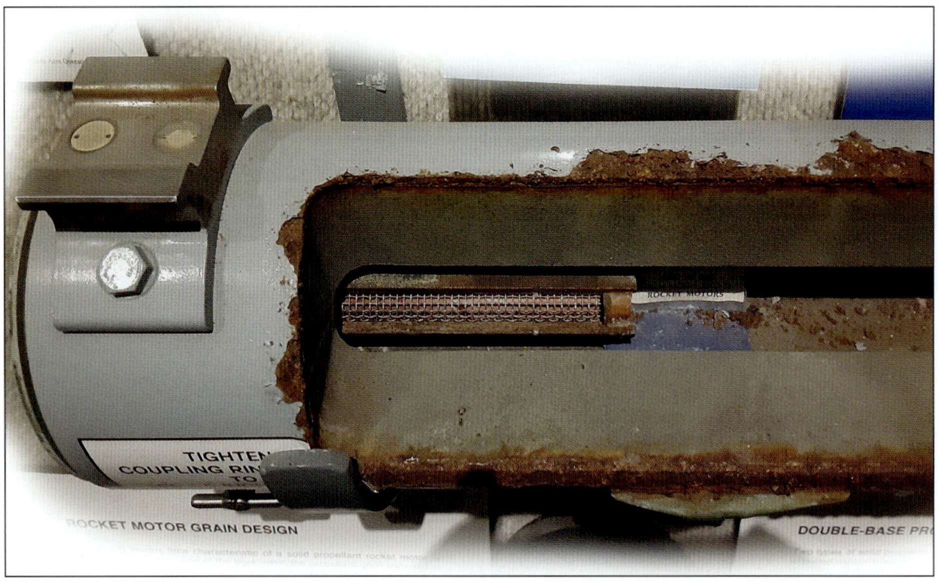

The igniter at the top of a Sidewinder missile motor solid propellant center perforation is shown in a display. One of the launcher hangers is also seen. (Author)

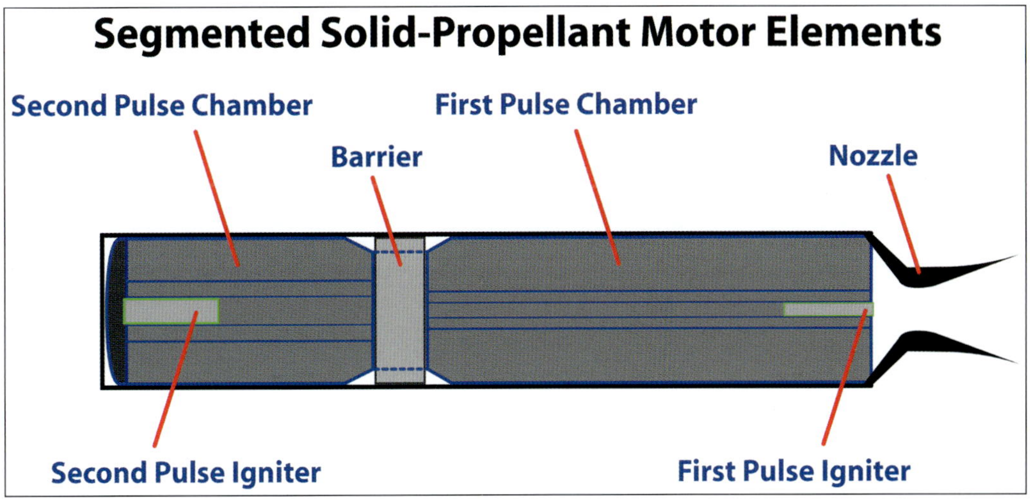

Figure 17: The segmented rocket motor shows how two individual grains are separated by a barrier. The grain just ahead of the nozzle is ignited initially as the first pulse, the boost phase. The drawing suggests this grain has a central core and is ignited appropriately but from the nozzle end. Ignition of the second pulse is delayed through a period of unpowered coasting until the optimal moment. It is then ignited for core burning to add energy and extend the flight time. (Drawing by author)

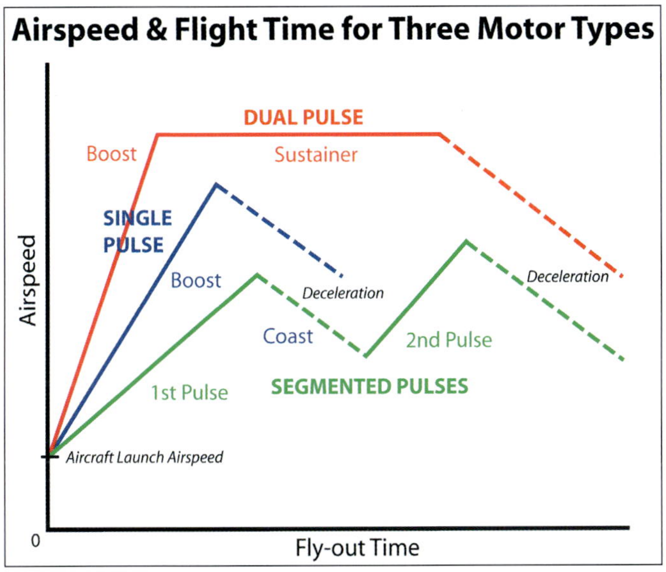

Figure 18: The airspeed to flight time relationship or three solid propellant motor types is depicted. The single pulse motor provides boost only and the AAM immediately begins to decelerate after burnout. The dual-pulse motor has a sustainer that burns immediately after the booster until burnout. The sustainer extends flight time and potentially also airspeed, all else remaining unchanged. The segmented pulse motor has a grain that burns first as the booster and the AAM coasts following its burnout. A second pulse is ignited at the optimal moment to add energy and flight time. The deceleration at the end of each represents drag dominating following fuel exhaustion and termination of thrust. The peak airspeeds and fly-out times for each are for illustration purposes only; they could be higher or lower depending on the motor. (Drawing by author)

This can cause irregular burn when ignited, burning through the case, and even explosion. Suspending the motor from points not indicated by stencils on the case can also promote invisible grain cracking. The same sort of non-destructive inspections as for the case can be utilized to search for flaws. There are typically temperature limits for storage of rounds, instructions for proper handling, and transport cases for the motor section to ensue against sharp impacts. Manufacturers struggled with small tolerable temperature ranges for many years, to the frustration of operators, until overcoming this problem. Some propellants are rubbery and can 'flow' over a long storage period, altering its initial internal design. This requires the missiles to be rotated occasionally during long-term storage. Some motors are replaced during a missiles' service with improved models.

Liquid Propellant

A rocket motor with liquid propellants typically requires an oxidizer and a fuel that are pumped to a mixing or combustion chamber ahead of a nozzle. However, liquid propellants are usually hazardous; toxic, corrosive, and harmful to exposed flesh. Their handling requires special equipment, careful storage and trained personnel. Loading of the propellants is typically performed off-aircraft and in a safe area should there be an accident. While some designs permit loaded missiles to be stored for long periods, they may need to be defueled occasionally. Such missiles also required some rounds to be sacrificed periodically for static firing to check that systems are not deteriorating in storage.

Among the earliest liquid propellants were hypergolic fuels, normally two components spontaneously combusting on mixing and so requiring no ignition source. The fuels were both toxic and corrosive, with any accidental mixing outside the combustion chamber potentially catastrophic. This demanded considerable transport and loading safety practices plus design measures. The extreme hazards of hypergolics meant that they were usually rejected and disappeared when more practical alternatives became available.

Missile rocket motors are stored for long periods and then expected to perform perfectly when ignited. To help ensure this outcome, a few are periodically fired on ground test stands, under strict safety protocols, to verify nominal operation without degradation. The restrained motor is ignited, and the thrust produced measured. (US Navy)

This example liquid fuel rocket motor is the Reaction Motors LR2-RM-2 unit for the Lark surface-to-air missile of the late 1940s. The mixing and combustion chamber is at left. The nozzle is not evident because the fuel was circulated in a jacket (shown) around the chamber and nozzle for cooling and pre-heating of the fuel. The 200lbf (0.9kN) motor used hypergolic propellants – aniline and red fuming nitric acid for the oxidizer. (Author)

Gelled Propellant

A hybrid is a gelled propellant that, when liquefied during operation, works as a liquid but, during storage and captive carriage, is like a solid. This can be the fuel, oxidizer, both, or a monopropellant. It also holds the promise of storing more energy than a standard liquid propellant to deliver greater missile performance. Acceleration alone may be sufficient to liquefy the gel to then be vaporized and sprayed rather than pumped.

Jet Fuel

The oxidizer in solid propellants can make up as much as 80 percent of the mass, and so eliminating that for an air-breathing motor promises more fuel for longer flight time and range plus the potential to throttle the thrust. A few missile types have employed some manner of jet propulsion, particularly if long range is sought since rockets burnout comparatively quickly. Given that the size must be quite small, motors with no moving parts are desired and so ramjets are most common (Figure 19).[7] One or more suitably designed inlets (or intakes) and an exhaust is necessary. Exceptionally high altitude means less oxygen in the atmosphere to support combustion and lower density of air likely not matching the inlet design. Consequently, inlets and nozzles can be of variable geometry to optimize performance for the flight conditions, but the design penalty is the added weight and complexity. Inlets are usually closed off until needed to reduce drag, a plug expelled prior to ramjet ignition. The propellant is some form of jet fuel (many suitable) in a tank along with a control valve and pump to pass the fuel to the combustion chamber. There it is vaporized via a series of injectors, mixed with the incoming air, and ignited by flame holders that ensure the spark is not blown out. A fuel control valve can serve to throttle the thrust. A ramjet can only operate at high velocity, and so some initial push off the aircraft to operating speed is necessary. This is usually achieved via a solid-propellant booster rocket. Thrust

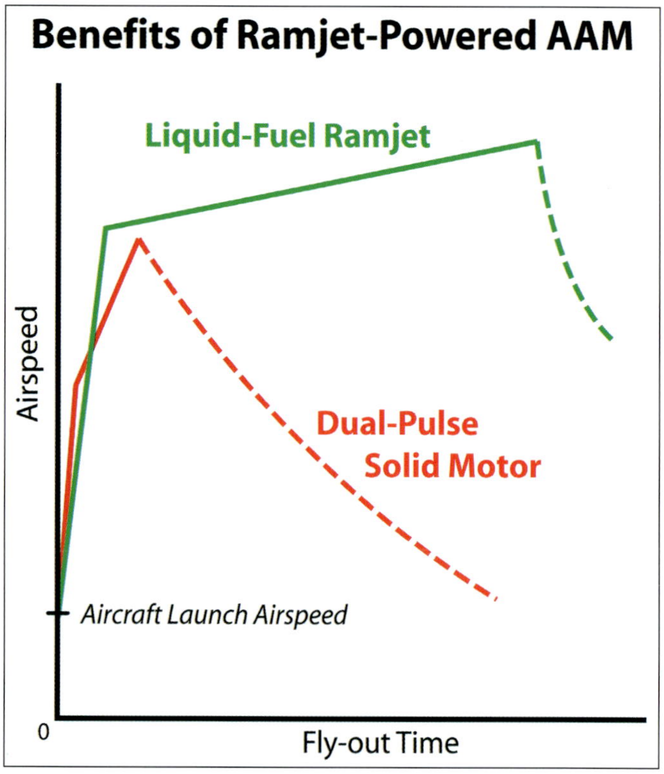

Figure 20: The advantages of a ramjet over a dual-pulse solid propellant motor is illustrated – expecting higher airspeed by virtue of the constant acceleration and longer flight time (equating to range) owing to longer burning of the fuel. The initial aircraft velocity is seldom high enough for ramjet ignition and so the missile must be boosted to engine operating airspeed by a rocket motor. The acceleration following that initial phase is by virtue of thrust increasing with airspeed. Subsequent deceleration with either motor follows fuel exhaustion and drag dominating.
The drag of a ramjet-powered design will likely be higher, given the protruding air intake, and so the deceleration is more rapid. The higher airspeed at ramjet cutoff can provide greater energy for endgame maneuvering. (Drawing by author)

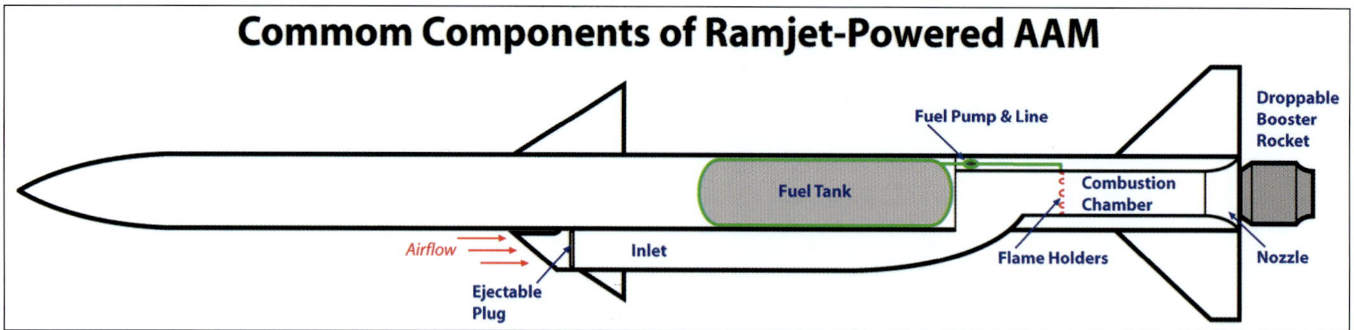

Figure 19: A simple drawing of a ramjet-powered AAM shows a configuration with a rocket at the rear to boost the round to ramjet operating speed. It falls away on exhaustion to expose the nozzle and permit the ramjet to begin operating. A plug in the single, fixed geometry inlet prevents the high drag of the open cavity when the ramjet is not operating. Fuel is injected into the combustion chamber and ignited via flame holders. (Drawing by author)

increases with velocity and so ramjets promise high supersonic flight and, with sufficient jet fuel, will operate much longer than any rocket for extended range (Figure 20). The typical skid-to-turn of a missile could be adverse for effective inlet operation, requiring roll-to-turn maneuvering.

By the 1980s work was being done with the integrated rocket-ramjet. Once the fuel of the solid-propellant booster is exhausted the motor casing serves as the ramjet combustion chamber. One complication is that the rocket motor nozzle may not be ideal for the ramjet and so it may be ejected to expose a more suitable second nozzle. More exotic is dispensing with liquid fuel and using a solid material that, when burned and the gases exhausted into the combustion chamber, serves as the fuel. A valve regulating the flow of these gases acts to throttle the thrust produced.

Flight Control

Fundamentally the missile is like any other object under stable flight through the air. It commonly features surfaces projecting from the main body to help sustain static and dynamic stability in three axes if motion (Figure 21). The ability to stabilize the missile in the presence of disturbances depends on placement of these surfaces about the missile's center of gravity (CG, the point about which the body balances statically), the positioning of the CG with respect to the centroid of aerodynamic forces acting on the body (the center-of-pressure, CP), and also flight conditions (primarily altitude and airspeed). The CP will move as shock waves form and shift across the body in transonic and supersonic flight, affecting stability and control. Commonly four fins and other stabilizing surfaces are employed and clocked at 45deg to the lift vector. (Some AAM, including the Sparrow, rotated after launch to assume this orientation.) The size of the surfaces must be such as to stabilize the projectile from the lowest speed of initial flight and in the least dense air anticipated (high altitude). As motors became more powerful, and initial velocity imparted to the round very high, the size of these surfaces could be reduced to save weight, reduce drag, and also shrink dimensions of the AAM aiding internal carriage. This is especially true if vectoring motor thrust (see later) assists in flight control. Some missiles have folding fins to reduce carriage volume, these popping out (usually under a spring force) and locking during launch.

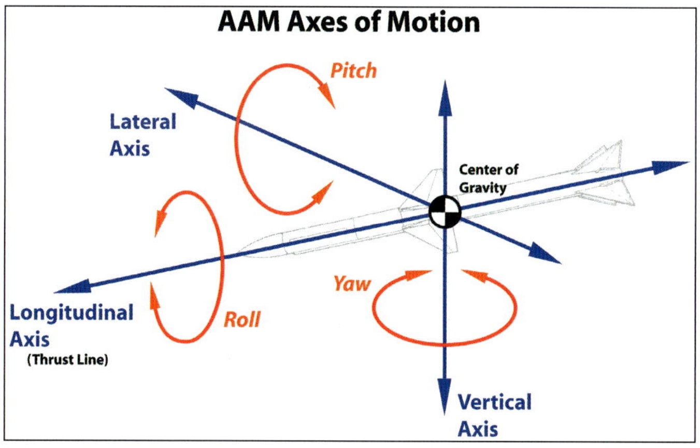

Figure 21: Shown are the three axes of motion, linear and rotational, for a missile in flight. These are the six degrees of freedom acting about the center of gravity. When disturbed in flight by, say, a gust the missile will react in one of these axes and the flight controls work to correct and to fly the missile per guidance commands. (Drawing by author)

Commonly, the missile's CG is ahead of the CP for static stability. Static stability implies the vehicle will right itself if disturbed in any axis. As mentioned, the CP moves with flight condition and the CG with propellant burn, and the flight control system would usually be required to deal with any anticipated relative position of the two. For a statically unstable vehicle, the CG is behind the CP. This would be like a tail-heavy arrow that would tumble out of control when shot. It is sometimes desirable to design the AAM to be statically neutrally stable (CP and CG coincident) or even unstable (CP ahead of the CG) in order to reduce aero surface area and to enhance maneuvering agility given that the control system does not have to overcome the normal stabilizing tendency to maneuver the projectile. The unstable design requires an active control system that works to prevent departure from controlled flight by sensing and countering any uncommanded motion within milliseconds.

Some of the control surfaces may be all-moving or may be fixed and contain displaceable surfaces like rudders and ailerons. Surface movement varies the pressure distribution on the surface to cause the missile to alter its flightpath (principally pitch and yaw) and possibly rotation (banking to turn and spin about the longitudinal axis aiding in stability). Their effectiveness or 'power' is a function of air density (less at high altitude) and velocity (decreasing following motor burnout).

In the first fraction of a second or so of flight, as the missile is accelerating from the velocity of the launch aircraft, the flight controls may be locked, not reacting to guidance commands, such that the AAM will not abruptly cross in front of and endanger the aircraft or another in a formation. The missile is flying a boresight ballistic path during this period. The surfaces are unlocked at a time the AAM is expected to be a safe distance ahead, Still, some pilots are startled by the round suddenly turning across the aircraft's flightpath. During that initial uncontrolled period the missile will tend to 'weathercock' into the relative wind if there is an airflow acting in the vertical or lateral axis from attitude of the aircraft, local flow around the aircraft, dynamics at separation such as a pitch rate, or just winds aloft. The flight controls, once active, must have the aerodynamic power to overcome this regardless of launch conditions. Driving the missile to elevated angle-of-attack (AoA) can also produce lift off the body (low pressure on top) or shed air vortices that can produce a motion in a certain axis. (The AoA is the angle between the missile longitudinal axis or the aero surface and the incident airflow.) This 'body lift' has been used to supplement classical aero surfaces or, if unintended, create forces the control system must overcome.

Thrust vectoring control (TVC) has also been employed for flight control by displacement of paddles or vanes in the motor exhaust stream to generate a force off the longitudinal axis of the missile (pitch and yaw). These demand paddle material that can withstand the extreme heat and pressure of the motor efflux for the duration of the burn. The buildup of expelled products of combustion onto the paddles as 'slag' is also to be avoided to ensure effective control. The presence of the paddles causes a disturbance in the exhaust flow and so a slight loss of thrust. Swiveling the nozzle is another TVC method. Attitude control via small reaction control thrusters, with action perpendicular to the longitudinal axis, can also serve to steer the AAM, as in spacecraft operating in vacuum. These can be valves venting stored pressurized gas or bleeding main motor exhaust gases, and one-use solid charges.[8] Such Divert Attitude Control Systems (DACS) or Attitude Control Motors (ACM) can particularly work during the initial phase of flight for a high off-boresight AAM that needs to perform very sharp maneuvers to engage a target well outside the forward quarter of the launch aircraft. However, very

The thrust vectoring nozzle of the AIM-9X is shown with the four paddles projecting into the motor's exhaust nozzle. These serve as supplemental flight control to the actuated fins shown, but only during motor burn. (Jim Rotramel collection)

The Attitude Control Motors (ACM) segment for a missile is shown in a shop environment. Each of the depressions would contain a small charge to emit high-velocity gases perpendicular to the body to affect motion in the opposite direction. In this way the vehicle can potentially be steered much more sharply than with aero surfaces and even TVC. (Lockheed Martin)

sharp maneuvers in the endgame for a hit-to-kill (HTK, eliminating the warhead and PF) has also been envisioned. The ejection of compressed gas into the exhaust plume to divert the flow has been used with large missiles but is generally impractical for AAM. The TVC is only effective during motor operation, typically a fraction of the entire flight time. Consequently, aerodynamic control has commonly been used together with TVC, the two means sometimes linked mechanically by use of the same actuators. For high off-boresight AAM, the use of TVC during the initial phase of flight, that may include abrupt turns towards the target, has been sufficient, endgame maneuvering then via the aero surfaces.

All of these functions require some manner of actuation (Figure 22) to displace the surfaces against the pressure of the airflow about the missile or the motor exhaust jet. These, in turn, require motive power such as electricity, pneumatic pressure, or hydraulics. Pneumatic pressure is usually via a pressured bottle of gas or a gas generator, released via a valve and moved via tubes. Hydraulics are typically the fastest acting but require a pump to move the fluid in tubes and, in turn, operated electrically or via pneumatics to circulate the heavy fluid from a reservoir. An alternative is a pressurized loop of trapped fluid. Power for electric motors come via batteries or a generator producing a current via a spinning dynamo. Electrical actuation has been the most desirable as it eliminates tubes and storage tanks for pneumatics and hydraulics. Electric actuators also generally operate faster than those alternatives. However, during the early decades of AAM development electric actuators were large and heavy, did not produce required torque, and actuation was a large draw on the current source. This changed in later decades such that electric actuation became the norm.

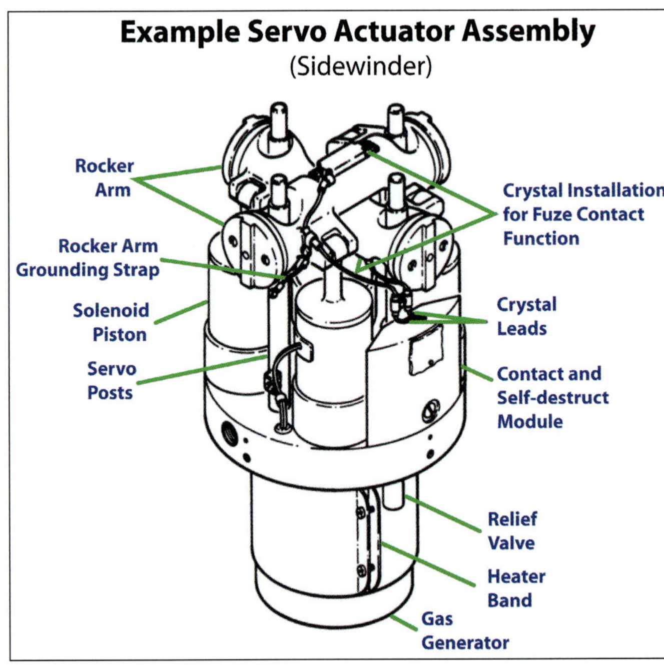

Figure 22: The elements of one model Sidewinder missile compact (5.0in/12.7cm diameter) servo control assembly for the four forward canards is illustrated. Stacked rocker arms to the canard surface attach points are actuated by pistons moved by pressured gas or electric solenoid. The canards moved in pairs. Later AAM used electrical actuators almost exclusively. (US Army, Modified by author)

For controlled free flight, control surfaces must remain effective from initial launch and acceleration, through tracking maneuvers, to the endgame and warhead detonation. This is despite missile attitudes and rates, airflow separation, airframe deformation from flight loads and maneuvering, altitude (air density), shock waves and wake encounters. As with any aero surfaces, there is concern that the airflow angle of the incident to the surface (AoA) grows so great that the flow does not remain attached but separates from the surface as shown in Figure 23. This means that the pressure distribution across the surface is greatly altered and may be ineffective for stability and control. This is called stall and can induce motion about the axes of the missile that may no longer be controllable. Especially where high maneuverability is required, equating to high AoA that may promote stall, some aerodynamic measures are likely required to delay stall, or other control means employed such as thrust vectoring. It is possible to place a fixed vane ahead of the surface to shed a vortex that energizes the airflow and helps to delay separation to higher AoA. This is also shown in Figure 23.

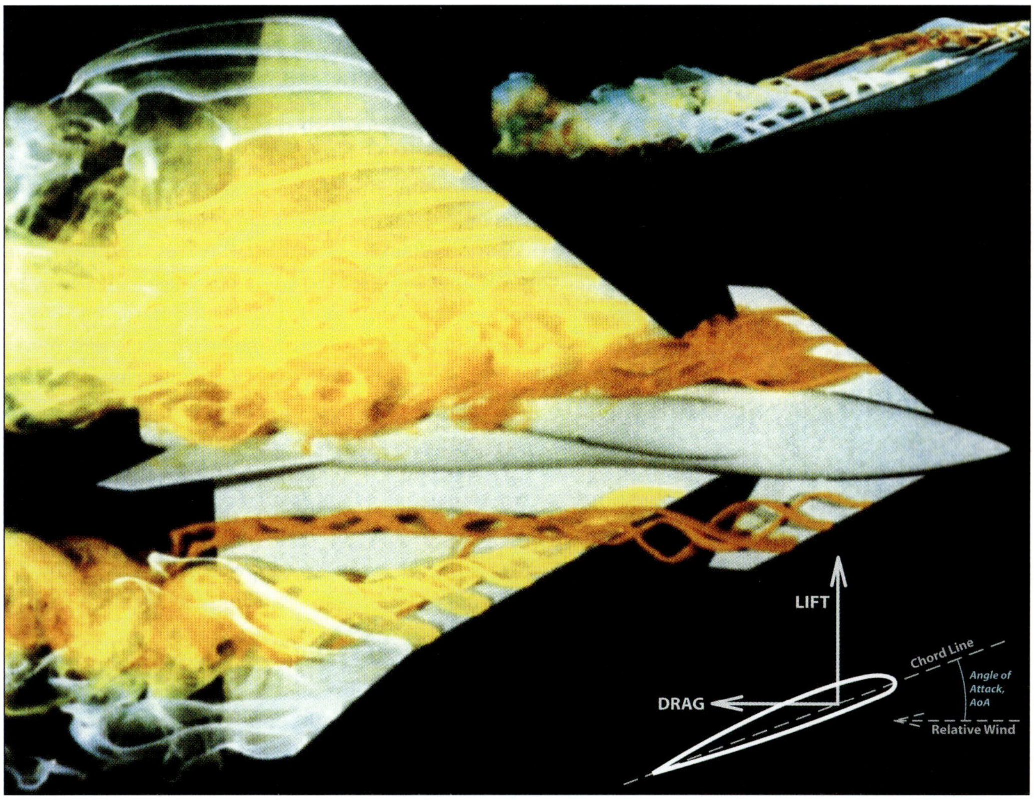

Figure 23: This photograph is of a fighter aircraft model in a water tunnel (like a wind tunnel but just a different fluid) with dyes injected to visualize the 'airflow.' The model is at high AoA (see inset drawing of a wing cross-section for a pictorial definition) and the wings are suffering flow separation (see side view at top right) and loss of lift. The port wing is essentially fully stalled (top view) and, if it were a missile control surface, its aerodynamic effectiveness would be nearly zero. The canard ahead of this wing is partially stalled and its separate flow is further disturbing the wing. The starboard wing has much less separated flow thanks to the vortices shed from the canard upstream energizing the flow. If this wing were a control surface, it would still be effective even at such a high AOA thanks to the influence of the upstream surface. (MBB, Drawing by author)

Autopilot

Stabilization is commonly via inertial measurement unit (IMU) measuring disturbances in the three linear and three angular axes of the missile presented in Figure 21. These have used a trio of gyroscopes measuring the angular rates and three accelerometers for the linear. A gyroscope is a spinning device whose rotating axis remains fixed in space despite motion disturbances. This initially used actual mechanical gyros that had to be spun-up prior to launch and installation had to be such that it remained free of the vehicle's dynamics. The device is built such that relative displacement from the initial stable condition in its rotational axis (precession) produces an electronic error signal. These have sometimes been called 'strap-down' systems, implying restraint in one or more motions such as pitch. The mechanical gyroscopes are only suitably accurate over relatively short time periods before errors accumulate (drift) undermining guidance accuracy. They were replaced during the 1970s and 1980s with solid-state devices like ring laser and fiber optic gyros. These solid-state gyros (no moving parts) use the Sagnac effect with light split to follow the same path in opposite directions. Under a rotational disturbance in the plane of the ring one beam requires more time than the other to traverse the path, and that difference (interference) can be scaled to angular velocity. Such devices possess much less drift and a very low rate of failure compared with a mechanical gyroscope. Accelerometers measure linear rate via inertia imparted to a mass that then alters a current proportional to that acceleration by either changing the length of a conductor via strain or compressing a semiconductor material. There is usually an assembly of three accelerometers, measuring linear rate in the three axis of vehicle motion from which velocity in any direction can be determined.

The output of the IMU, a combination of gyro angular rate motion and accelerometers linear rate motion, is used by the autopilot to correct for unintended motion via commanded control surface movement and then compares the resultant motion via feedback for further correction. Altitude and airspeed readings (via the measurement of atmospheric pressure) and heading reference (with respect to the Earth's magnetic north) may also be utilized. However, these are usually reserved for very long-range missiles as most AAM guide to the target regardless of such quantities. Similarly, such long-range missile may correct for IMU drift by periodic reference to a Global Positioning System (GPS, or equivalent satellite positioning system) location determination.[9] This is called GPS-aiding.

Guidance/navigation signals (see next) are passed to the autopilot which affects the commanded change to a new course. This is done by determining which control surfaces to move and by how much, then sensing when the course change has been accomplished (error driven to zero). Some early missiles had only one pair of control surfaces, like two opposing canards that were actuated while the other pair was fixed. This requires the autopilot to roll the missile as necessary to affect a change in an axis – rolling level to pitch and 90deg to yaw – or 'twist to steer.' The airframe would usually be rolling during the entire flight and then the surfaces only moved at the correct point in the roll to affect the steering change. (Some missiles have canted aero surfaces to naturally induce a roll during forward flight for inherent stabilization. This demands the same autopilot control scheme to command course changes.) Such control schemes produce what looks like an erratic flightpath. Should the missile deviate from the commanded flightpath, perhaps by the missile being disturbed by air gusts like wind shear or wake encounter, the autopilot works to correct back. Additionally, the same control deflection for this task in dense air at low altitude would be insufficient at high altitude, and the autopilot must compensate appropriately. Guidance, navigation, and autopilot functions are frequently performed by the same electronic components of the missile or shared components in a federated system.

Guidance

Navigation algorithms (many available), generating commands to be implemented by the autopilot, driving the missile to an intercept course with the target based on input from the seeker, usually rate of change in line-of-sight (LOS, Figure 2) from the seeker tracking mechanism (see later). This track of the target permits prediction of its future position. That was previously only possible with radar, but it was eventually possible to equip electro-optically guided missiles with this processing capacity. For head-on engagements, where closing rates are much higher than tail chases, the gain must be set very high such that corrections are more frequent and finer. This, however, may cause excessive motion for other aspect attacks (snaking flightpath) that must be compensated for (variable gain). The early missiles also used a 'bang-bang' or modestly modulated guidance, with sharp and possibly full control surface deflections for every input – three positions of full deflection in either direction and neutral – that made for harsh corrections and so an erratic flightpath. This was especially true for rolling missiles employing only one opposing pair of control surfaces.

Figure 24 shows the four possible intercept trajectories. An optimal intercept requires a missile flying substantially faster than the target to ensure against it becoming a pure pursuit ('curve of pursuit'). Pure pursuit means that the guidance systems keep the flightpath of the missile (or velocity vector) aimed at the target for the entire time of flight. The AAM is in a continuous turn of ever-decreasing radius and, in the endgame, may lack the energy to complete the final tight turn. Even worse is the lag pursuit with the AAM fired with LOS behind the target (lag rather than lead). The round has to turn sharply in an effort to establish lead and may overshoot the target's flight path owing to excessive energy, then correct back to enter what may become a pure pursuit in an especially inefficient manner. Leading the target visually or via a radar-return calculation can produce the more efficient lead pursuit flightpath that still ends in a tail chase with potentially sharp and energy-draining endgame maneuvering. The most efficient intercept guidance – less flight time and little terminal phase maneuvering – is the lead collision. This requires the target to be tracked such that its future position can be calculated and the AAM guided to the point in space to meet the target. Ideally this would be a straight line for a non-maneuver target. More normally, continual calculations and adjustments to the AAM's flightpath are necessary.

The lead-collision intercept has been frequently done with proportional navigation computations, a.k.a. convergence. (This is a guidance technique and not navigation by the literal meaning.) Rate of change of LOS to the target is matched proportionally by a missile rate of turn until the rate of change is driven to zero. This establishes a lead that, if this flightpath is maintained – the rate of change of the angle remaining unchanged – produces a straight flightpath to a lead-collision intercept (Figure 25). The seeker look angle or LOS is notionally constant and always off the longitudinal axis of the missile. In actuality, missile speed varies with motor performance or burnout then deceleration under aerodynamic drag, and the target may maneuver, so calculations are constantly updated and the flightpath corrected. This can produce a varying flightpath such as shown in Figure 26. From a boresight shot directly at the target the missile immediately turns right to arrest perceived target LOS drift and establish the lead. As the missile accelerates under boost, less of a lead angle is required and it turns left to maintain the constant

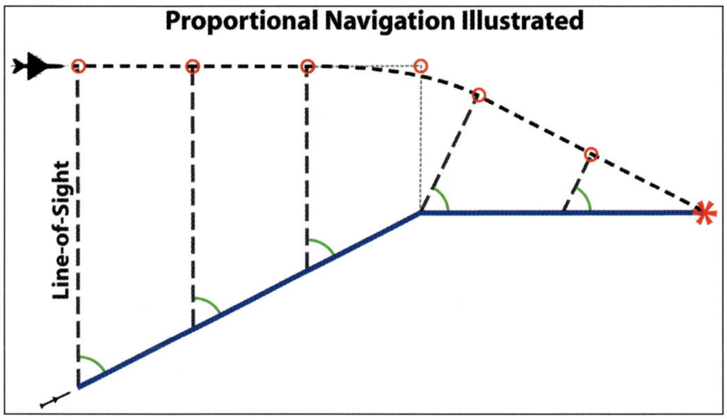

Figure 25. This illustration of proportional navigation shows that the angle between the missile' flightpath and its seeker line-of-sight to the target remains unchanged (rate of change is zero) before the adversary makes the turn and after, though a different angle is established following the maneuver. (Drawing by author)

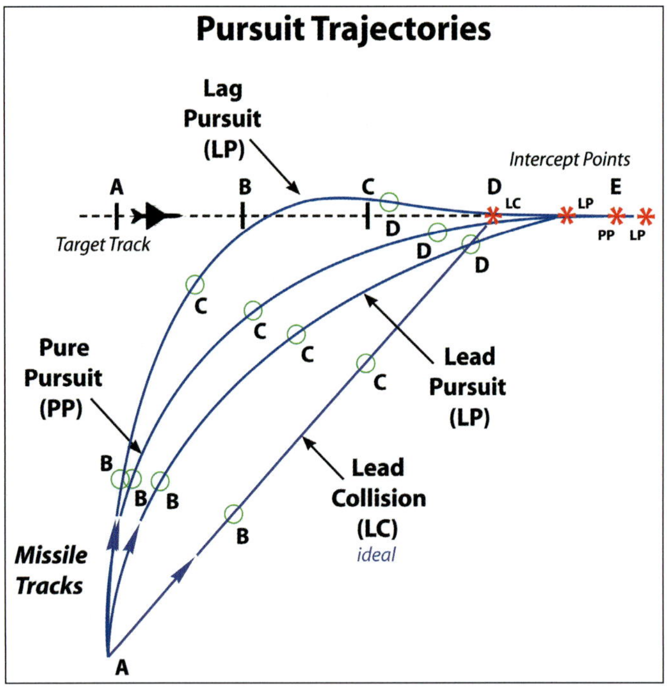

Figure 24. The four possible pursuit trajectories for an AAM to intercept an aerial target are depicted. It presumes a missile velocity approximately 1.5 times that of the target that is following a straight flightpath. The unit of time intervals could be any number of seconds. (Drawing by author)

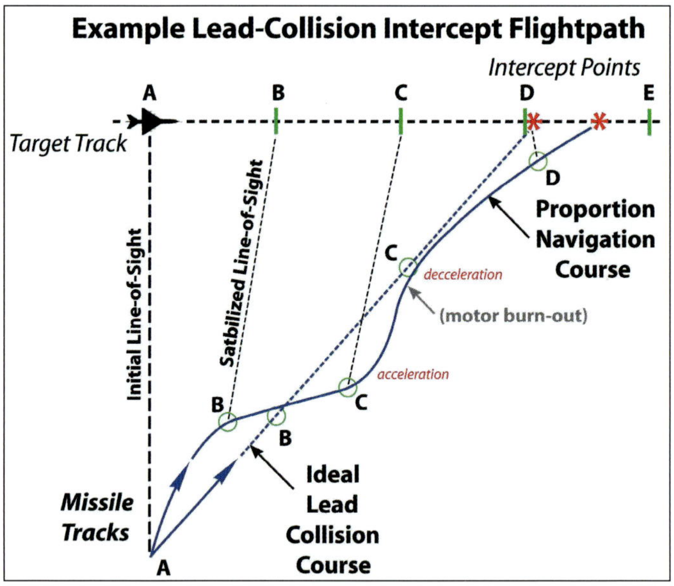

Figure 26. Depicted is the variable flightpath of an optically guided missile seeking a lead collision intercept via proportional navigation given AAM speed changes during its flight. This is compared with an ideal straight flightpath for a constant speed intercept that required less time to affect the intercept. Actual intercepts also occur in 3-D space rather than this 2-D representation. (Drawing by author)

The flightpath of an AIM-9 illustrates the effects of proportional navigation with its serpentine smoke trail. The launch aircraft is an F-100D and the target was likely a parachute-suspended flare. (Albert Grandolini collection)

LOS. After the thrust is expended and speed begins to bleed, another lead angle is necessary, established with a right turn. Add target maneuvering with changing aspect angle and the flightpath can become even more convoluted. A pilot observing this as a snaking smoke trail may assume the AAM has malfunctioned and fire another before the first has clearly hit or missed. The update rate for guidance corrections is dependent on seeker scan rate; the higher the rate the more refined the guidance and target tracking. How hard the missile turns to maintain the optimal flightpath is established in the autopilot as a constant of proportionality or navigation ratio (K factor, Figure 27). Lateral acceleration for a turn determined by K is proportional to LOS change, and hence the name of the guidance technique. A K of 1 produces an ever-tightening turn to convergence. This is lessened by greater K factors via earlier turns during the fly-out. Closing velocity determined via Doppler (see later), if available from the seeker, can be employed to optimize this guidance solution.

More advanced missiles also incorporate perceived target range, velocity, acceleration and time-to-intercept in the guidance calculations for ever-more efficient flightpaths. A less snaky and inefficient path can be created by altering the weighting the LOS drift is given in the navigation calculation during the flight (variable gain). This can also reduce initial low-speed maneuver to help ensure against loss of lock with an initial low or fixed gain and improve terminal accuracy with other gains. The radar homing AAM has the advantage of determining range to the target and closure rate in addition to LOS. Modern missiles with considerable digital computing power can offer many alternative navigation solutions depending on varying scenarios.

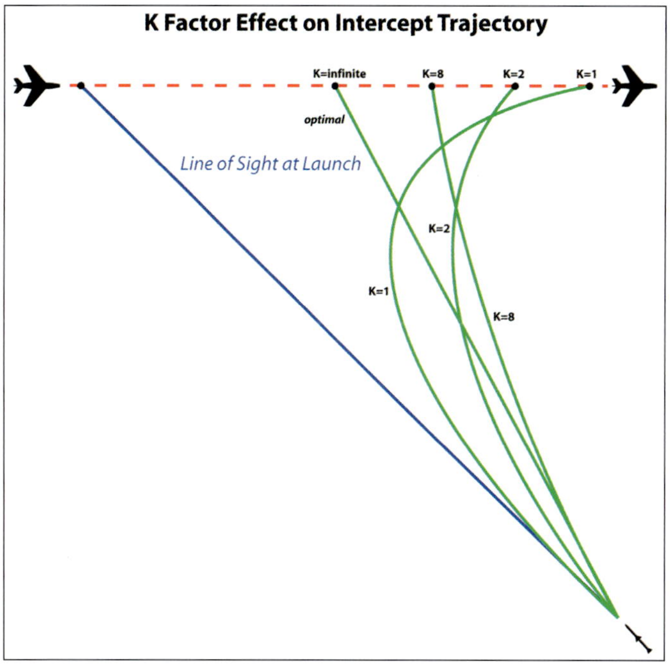

Figure 27. How hard the missile turns to achieve the intercept is a function of the K factor employed in the guidance system. As illustrated here, a K factor of 1 results in a high-g turn and a tail chase. A K of 2 produces a circular flightpath. Proportional navigation for a lead intercept requires higher values with varying turn radii and flight time, with more maneuvering at the initial stage of the flight, to an optimal requiring no turn (K = ∞ and so unattainable). (Drawing by author)

Power Generation

Electrical power is used to energize electronics but also pumps for pneumatics and hydraulics. Prior to firing, the missile receives electrical power from the launch aircraft. Just prior to firing the AAM's internal power source is brought on line and its output checked. There is generally a feature that aborts the firing if this fails to meet set criterion.

An airstream-driven propeller for a power generator was quickly seen as impractical for a supersonic missile. Other forms of generators requiring fuel and air for combustion were similarly rejected for weight and safety concerns plus the fuel consumed in powering them up prior to launch. A pressurized gas slowly released from a bottle to turn a generator consumes considerable volume. A slow-burning pyrotechnic to produce the pressurized gas turning a small turbine generator or alternator has been used successfully. That is a one-shot affair in that, if the missile is not launched it cannot be selected again during the mission. Batteries are the most practical source, but the initial lead-acid units were heavy, needed to be serviced and recharged frequently and occasionally replaced. Storing and handling electrolytes in a combat environment is hazardous, and ensuring suitable power to recharge the unit is another burden.

The AIM-7E's Electrical Power Unit is seen in a display. It uses a powder charge to spin up a turbine that, in turn, rotated an alternator to generate the current. It replaced a service-intensive lead-acid battery. However, the 8.0in (20.3cm) diameter section remained heavy and with many hand-assembled components. (Author)

By the late twentieth century the thermal battery became the desired stored electrical energy source in AAM owing to high energy density and long 'shelf-life.' These are generally activated by a pyrotechnic to melt an electrolyte that then releases current along with heat. Thermal batteries are not reusable and there can be some delay from activation to full current. They are ideal for AAM since they only operate for a few minutes before destruction.

Whatever the original electrical power source, the need to convert to alternating current (AC) and direct current (DC) at various voltages and currents demand additional electrical system components.

Arming

This is the function that readies the missile for firing such that the aircrew need only push a button or pull a trigger to launch. The arming may be an intermediate pull of the trigger following by full depression to launch, a separate switch, or selection on a multifunction display via bezel button or virtual button on a touch-sensitive screen, or even a voice command. In most cases the arming must be followed by some indication from the missile that it has passed some electronic self-test or has acquired the target, such indications (visual or aurally, or both) necessary prior to firing for confidence the shot will be successful – recall Figure 12. The first step in arming is commonly to move a Master Arm switch to ON, permitting selection of any available weapon.

There is also a ground crew arming of the missile for firing prior to takeoff as detailed in Chapter 4. The warhead fuze is also armed at a certain point in the missile fly-out as detailed in the next section.

Fuzing

The warhead is detonated at the right moment via activation of a fuze. An AAM fuze is generally some sensing mechanism to judge proximity to or impact (contact) with the target. The former is termed a proximity ('prox') fuze or PF. This can be a passive device (influence fuze) or active device like a radar or laser.[10] A near-miss is sufficient with a PF if within the sensing range that, in turn, is associated with the warhead design lethal radius (next section). The proximity fuze activates when a sensed phenomenon passes a set threshold, the minimum range cutoff. The physical time from the cutoff through explosive fragmentation dispersion may be critical for very brief passages through the lethality window based on closing speed and crossing angles. With encounter times of 250–1,200 milliseconds, a few hundreds of a millisecond of fuzing latency can make the difference between a kill and mere damage. Such short time periods also challenge the ability to replicate such conditions in the lab or a test range to optimize the system without expending many missiles – and such flight trials would be in the final stages of development process when unfortunate discoveries can be very disruptive to schedule. Even a close approach may not provide the necessary conditions to trip the fuze depending on many factors associated with the nature of the PF, the target and crossing track.

For safety purposes the fuze may only be enabled after firing to prevent premature warhead detonation endangering the launch aircraft via stray electrical charge or other malfunction. This can be accomplished by such a device such as an inertia switch – acceleration closing an electrical contact.

The sensed phenomenon for passive fuzing may be simply an abrupt change in LOS as measured by the seeker, indicating close passage to the target, and so not a dedicated missile element or function. The most common sensor-based passive fuze has used infrared detectors – measured IR emissions reaching a peak and then beginning to decay, indicating closest approach. There is no range determination and it has the same potential for confusion as the AAM's main IR detector (see later). Typically, four or more detectors arrayed around the missile's body circumference is used to ensure detection of the target. Another passive method that has seen some use is sensing the disturbance of the aircraft's metallic mass in the magnetic field of the Earth. As aircraft have little iron in their construction and so the effect on the magnetic field is exceptionally small, this method is not favored.

An active fuze typically employs Doppler Shift as a means of determining how close the round is passing the target and compares this with the threshold. Doppler Shift is the apparent change in frequency of a wave phenomenon owing to compression of waves (Figures 28 and 29) given relative motion of the source versus a moving reflective object. This can be used to determine range and velocity of the moving object via the waveform reflected off the target relative to the transmitted signal. Once the range falls within the minimum cutoff the fuze is activated. In some cases, a time delay is built in to permit the AAM to move along the target aircraft (assuming a rear or front aspect intercept) and then detonate opposite more critical areas with greater lethal effect. This bias becomes problematic when approaching from other aspects. The all-aspect AAM led to development of adaptive fuzes in which the engagement scenario, as assessed by various intersect parameters through onboard computer processing, determines the timing delay.

Active proximity fuzes have used RF emissions, commonly continuous wave radar (CW, see later). The fuze emissions are generally from two antennas on opposite sides of the AAM forebody, conformal with the outer skin and possibly not readily evident. The two receive antennas are usually clocked 90deg from the transmit units. This implies an expectation that the wide transmit antenna main lobes (Figure 10) are emitted such as to sufficiently cover a full half circle, or nearly so. As with any RF device, it is subject to jamming and glint that may prevent suitable operation.

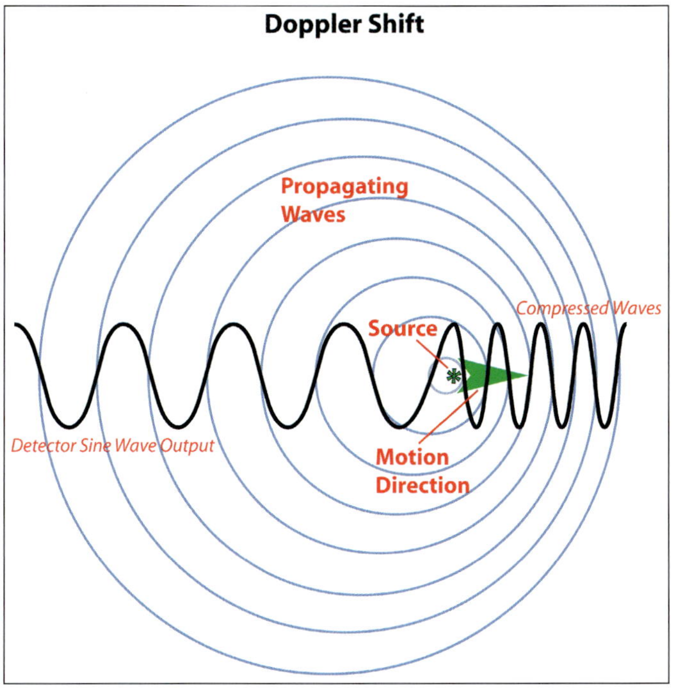

Figure 28. This drawing depicts the Doppler Effect or Doppler Shift phenomenon from the perspective of a fixed detector. The waves (sound or electromagnetic) generated by a moving source will see the waves ahead of the source compressed compared with those behind. This shift is measured as sine waves of different wave lengths and frequencies. The drawing is a two-dimensional representation of a three-dimension physical reality. (Drawing by author)

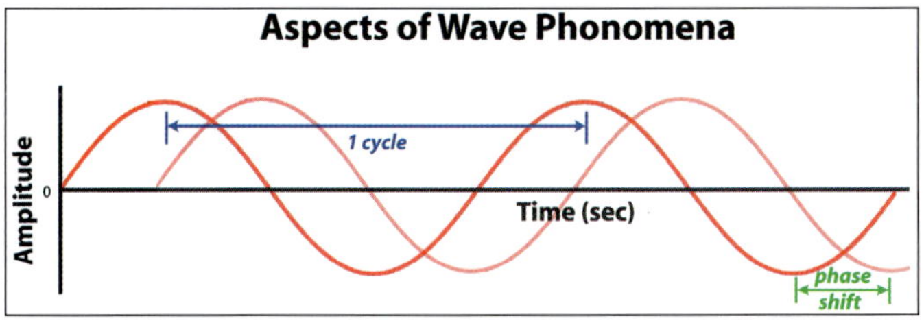

Figure 29. A wave phenomenon is depicted with a simple sine wave, though it can be other shapes such as a square wave. It also need not oscillate about zero amplitude or be sustained (undamped). Frequency of the waveform is the number of cycles (peak to peak, the wavelength) that occur in a second, measured as Hertz. A change in the waveform may define a phase shift in comparison to the original, as illustrated. (Drawing by author)

A laser proximity fuze is a compact and complex bit of hardware as suggested by this photograph of an AIM-9L's DSU-15 Target-Detecting Device (TDD). The laser light emitter is on the left and the reflected light sensor on the right. (Jim Rotramel collection)

When compact, solid-state lasers became available in small packages these were adopted in PFs because of their advantage in being almost immune to countermeasures. The laser light is emitted from a pulsed source, usually a light-emitting diode shining out a window in the side of the AAM. This wave of light is spread out via optics into a narrow fan beam of 90deg or more width (Figure 30). The reflected light waveform is received through the same window or another. Three or four laser/sensor pairs can be packed into small diameters weighing on the order of 20lb (9kg). Twice the number of emitters allows a pattern sent forward at perhaps 45deg to make initial detection and a second set to emit perpendicular to the AAM body to trip the fuze with the round closer to the middle of the target aircraft. If the target passes through the first beam without being detected by the second, it is assumed to be a small target and fuzing is commanded.

The PF is usually complemented by an impact or contact fuze (a.k.a. direct action, DA). The DA fuze may be tripped by an inertia switch, closed by abrupt deceleration, or a contact mechanism such as a broken circuit (severed conductor) or stretched conductor altering resistance. The impact fuze is seldom sufficient as a direct hit has proven rare. However, this is still commonly incorporated owing to its simplicity and in the event the PF fails, or the engagement is such that it is not tripped. For example, should an IRH missile go straight at the exhaust of an aircraft from directly astern and 'goes up the tailpipe', the proximity fuzing minimum range cutoff criteria may not be satisfied. Such would make a contact fuze highly desirable. A variant of the contact fuze, and sometimes used in conjunction, is the grazing fuze. This senses a change in velocity or direction of flight beyond a specified threshold.

Many missiles have a timer that detonates the warhead after a certain flight time corresponding to when the AAM would have entered a ballistic descent and so clearly missed the target. This serves as a self-destruct feature to help ensure against it homing on an unintended target (and possibly friendly) during its descent. It also ensures that the remains of the round will not fall into enemy hands if reaching the ground intact.

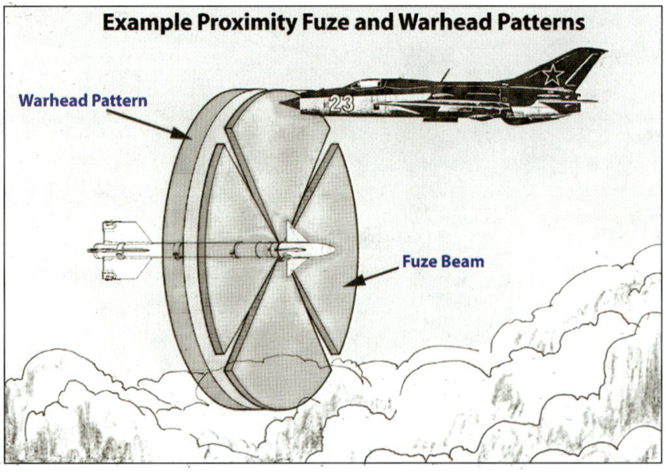

Figure 30. Depicted is the notional four-sector fuze sensor pattern – a beam suggesting active means such as lasers or radar and figuratively the sensing pattern of passive fuzes such as IR. The warhead pattern also suggests an annular shaped charge (see later) with the principal force dispersed in an annular pattern and possibly pushing out fragments or an expanding metal ring. (US Navy. Modified by author)

In this picture a fuze assembly is mounted atop the moving platform as it passes a model of a target aircraft (MiG-21 in this case) during a test in a US Navy facility. The model can be made to simulate radar reflectivity, IR emissions or other phenomena. The encounter between the fuze and the target can be conducted at many angles, distances and speeds with results recorded to assessment whether the fuze meets specifications. (US Navy)

Warhead

An AAM striking the aircraft target may damage or disable it, but it is more effective if the explosive warhead detonates inside or nearby. This is especially true in that a physical hit is rare. Missiles that have struck an aircraft but without warhead detonation have been known to pass through the structure without destructive damage. The ideal is to destroy the aircraft rather than allow it to be withdrawn for repair to fight another day. Warheads have also detonated with so little damage this is only discovered after the flight. Much effort goes into ensuring an AAM intercept ends with lethal effect.

The warhead explosive is set off by an electrically activated pyrotechnic initiator (detonator) as commanded by the fuze, as described previously. A safe and arm (SAFE-ARM) function prevents or allows the fuze signal reaching the detonator explosive. This can be a pin inserted and removed by ground crew or a rotation switch moved by same. The detonator ignites a lead that, in turn, ignites a booster that effectively ignites the entirety of the main charge. Another safety design is an inertia mechanism orienting the lead to complete the 'explosive train', this coming with the abrupt acceleration at missile launch and so preventing the main charge from detonating prematurely. These features are depicted in Figure 31.

The explosive blast generates a super-heated pressure wave, but with intensity decaying inversely proportional to the cube of the distance (radii, R) from the blast center ($1/R^3$). The pressure in the wave falls below atmospheric pressure before rising back to that ambient condition. This change from a compression to suction wave can further stress the target aircraft airframe, with internal pressures

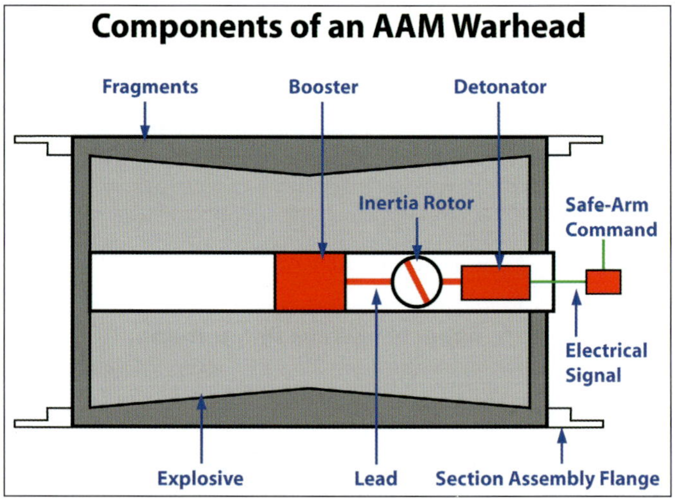

Figure 31. The basic components of the warhead section of an AAM are depicted. Electrical initiation of the detonator ignites the lead that, in turn, ignites the booster to set off the main charge in what is termed the explosive train. The inertia rotor element of the lead (actually, the entire lead) is rotated by a mass and spring tripped by acceleration of the missile at launch as a safety feature. A shaped charge explosive is shown, configured for annular blast. (Drawing by author)

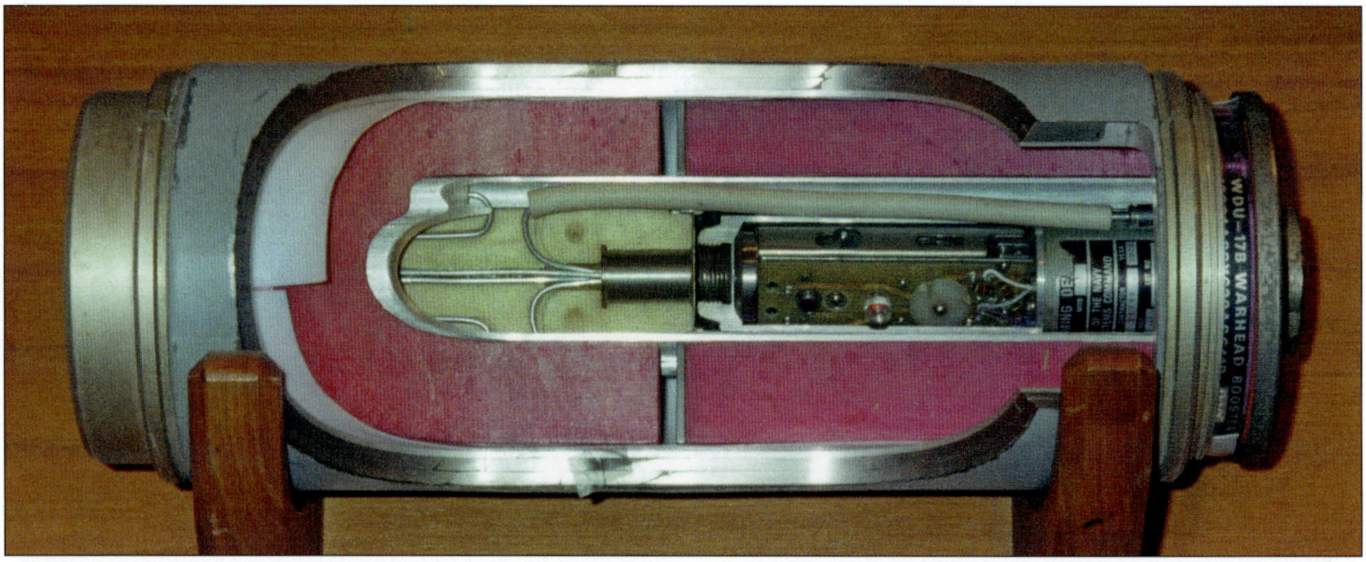

The WDU-17B warhead for the AIM-9L missile is shown in a cutaway display. Some of the internal elements can be compared with the associated drawing on this page. This was not a shaped charge unit and instead employed the ribbed liner between the explosive and the casing to score the explosive surface into tiny, shaped charges to produce suitable fragmentation. (Jim Rotramel collection)

within its volume causing it to rupture. The intensity of the wave is also dependent on the ambient air density such that destructive effectiveness is much reduced as altitude increases. This makes warhead fragmentation highly desirable.

The warhead is commonly designed for the high explosive to displace hardened metal fragments encircling the explosive as a case that may also serve as the outer surface of the warhead section. Fracturing of the case into multiple elements may serve as fragmentation, or elements like ball bearings might be incorporated. The fragments initially lag the pressure wave but soon overtake and pass through it to travel farther than the pressure front. Fragments also lose velocity over time, and so penetrating potential, as they spread out ($1/R^2$). The heavier fragments travel farther than the lighter, and larger fragments cause greater damage. However, a tradeoff in added mass of higher density fragments versus warhead weight must be made.

Fragments are typically displaced in a 60–90deg annular spread, though with less density of fragments off the perpendicular path from the warhead. Losing penetrating potential with distance traveled means that detonation beyond a certain warhead radius may just leave non-lethal holes for the aircrew to be heroically photographed beside after recovering to base. The velocity of the displaced fragments has a forward component as a function of missile velocity as well as perpendicular, and this reduces their effectiveness in a parallel pass intercept. Warhead radius of effectiveness must be considered with fuzing and possible miss distances. If large miss distances are likely, then a larger warhead would probably be adopted. Conversely, if guidance is more accurate, a small warhead may be sufficient.

Lethality can be enhanced by choice of fragmentation design (Table 6). The metal case may be pre-scored to ensure that it fractures into many small, high-velocity fragments. Explosive pellets or bomblets have also been employed, sometimes incorporating incendiary material, helping to increase lethality provided penetration is achieved. An annular blast fragmentation (ABF) warhead creates a ring of expanding shards to slice through the target given that there is high potential the missile will pass beside the enemy aircraft if it does not impact. This employs a shaped charge to impart the specific direction to the fragments, especially to the side, and so reducing the lower effectiveness of the forward velocity vector. One favored type of ABF is the continuous rod warhead in which linked rods make up the case around the explosive (Figure 32). This expands as a ring under blast pressure like a flexible metal watchband to more effectively cut through an airframe and its systems. The ring of side-by-side rods around the explosive are welded together at alternating ends such that, after reaching its full diameter, the links break into individual tumbling fragments. A Selectively Aimable Warhead (SAW, a.k.a. focused and directable) or configured blast fragmentation warhead uses means to initiate the explosive in a specific direction, as determined by fuze detection of the target's location relative to the missile. This maximizes fragmentation in the detected direction of the target for optimal lethality. One means is to have multiple detonators and setting off one of these creates the equivalent of a conical shaped charge. When the other detonators are set off a fraction of a second later the highest velocity fragments are then dispersed in the direction of the first detonator (Figure 33).

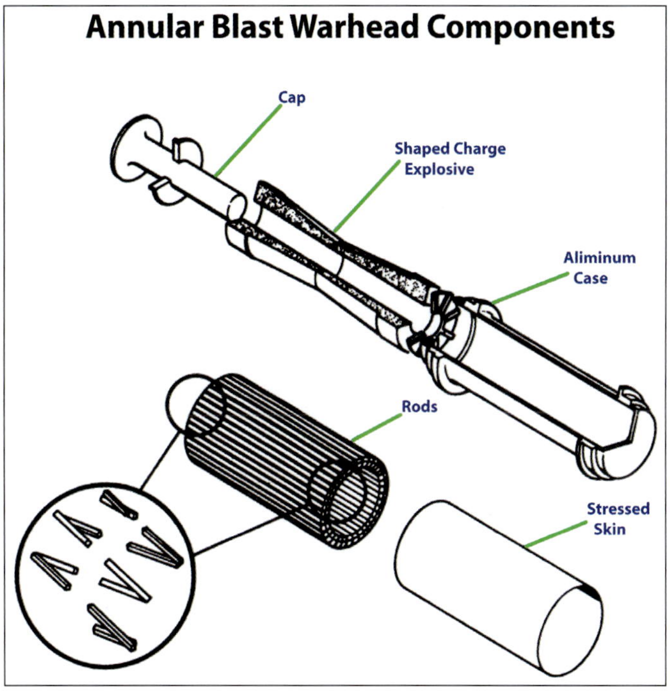

Figure 32. Depicted are the elements of an annular blast fragmentation warhead with a shaped charge explosive to create the expanding ring pressure wave plus displacing the surrounding continuous rod sheath. (US Army modified by author)

A test of an AAM warhead is shown in high-speed camera frames in which the camera has automatically adjusted for the brightness of the image based on the flash. The warhead is mounted vertically within the ring of witness panels and fragment velocity monitors. The initial blast is followed by dispersion of fragments. The witness panels, a staggered distance from the blast, records fragment impacts. The angled annular spread of the fragments is evident. (US Navy)

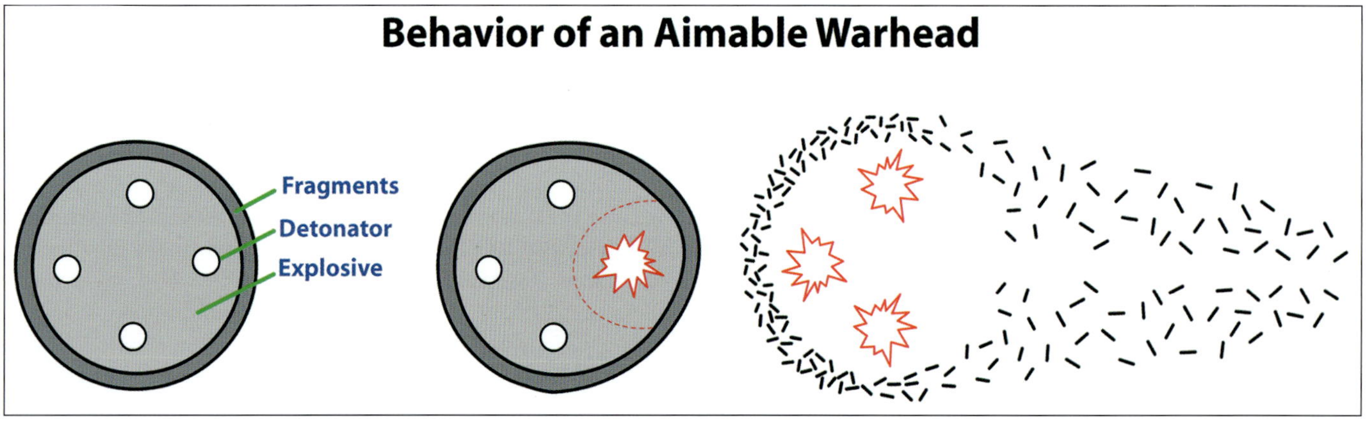

Figure 33. This shows the fundamental behavior of a 'smart' aimable warhead. It has multiple detonators with one set off initially in the direction of the target as sensed by the fuze. This creates the equivalent of a conical shaped charge such that when the remaining detonators are triggered the highest velocity fragments are propelled in the direction of the initial detonator. (Drawing by author)

A US Navy test on 15 August 1979 of an AIM-9 missile warhead has it detonating just above an F-89 on the test range as part of a lethality test. The central blast and expanding ring of high-velocity fragments are evident. The annular character of the fragmentation distribution can be seen by the impacts across the wingspan. The artifacts in the image, evident in the sky, are reflections off a protective transparency in front of the camera. (National Naval Aviation Museum)

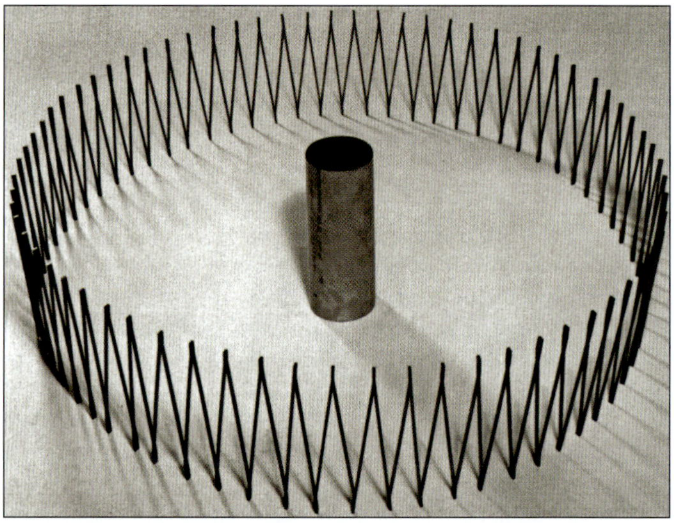

The continuous rod warhead fragmentation device is shown in its compact form (center) like a tube, as it would be positioned around the shape charge explosive, and partially spread out showing how alternate ends of the rods are welded together. (US Navy)

Table 6: Warhead Fragmentation Characteristics[11]

Warhead Type	Fragment Type	Fragment Velocity (fps)	Result	Notes
Fragmentation	many small fragments	6,000–8,000	penetrates internal components	wide 60–90deg spread of fragments
Annular Blast Fragmentation	discrete lightweight rods	6,000–8,000	cuts structural members	narrow spread, effective against high-velocity targets
Continuous Rod	joined rods expanding to ring	4,000–5,000	cuts surface and structural members	requires padding reducing explosive volume, narrow effect, weak against high-velocity targets
Aimable	discrete rods	>8,000 in aimed direction	cuts structural members	fuze must indicate target direction/aspect

In the early 1950s there was still concern with the ability of interceptors to clear the sky of enemy bombers preparing to deliver nuclear bombs. Using a nuclear warhead on an AAM would make a direct hit or close miss unnecessary. A nuclear detonation is so powerful that it is likely to damage or destroy targets within a large radius in addition to irradiating and likely blinding the aircrew plus impairing the aircraft's electronics with an EM pulse. These might readily wipe out formations of bombers. To avoid the crew of the launching aircraft suffering the same fate, they had to execute a hard 180deg turn and exit the area at maximum airspeed once the AAM was away. The potential for employing nuclear weapons was more readily considered in the 1950s and 1960s. There were thousands of these weapons of various sizes and yields in the inventory of the USA and USSR. They were employed in artillery shells, demolition charges, and rockets. The idea of using them as anti-aircraft weapons came easily. In the USA it was many years developing a miniaturized nuclear warhead suitable for such applications and, given the great secrecy surrounding the work, performed by the Atomic Energy Commission (AEC), Department of Defense (DoD) contractors had difficulty developing delivery systems given meager interface data.

The Soviet MiG-17PFU was a model of the capable interceptor designed to carry four RS-1U beam riding AAM. Dating from the mid-1950s, the jet had a Izumrud RP-2 S-band radar antenna in the inlet bullet fairing to locate a target and then guide the radar beam riding missile. This first-generation technique was very limited and the short-range RS-1U was rushed into service before being fully perfected in order to meet threats from American bombers. The weapon pylon had a long probe projecting ahead permitting a lanyard from it to be attached to a cap over the ram air turbine at the front of the missile that powered the AAM's radar receiver. On launch the cap was snatched away. It all made for a heavy and draggy installation. (Artwork by Tom Cooper)

A Jordanian F-104A shows the wingtip launcher with a first-generation AIM-9B. These displaced the fuel tanks typically installed at that station and so degraded range and endurance for the aircraft with a modest quantity of internal fuel. This is an example of a fighter not initially designed to accommodate AAM subsequently modified, imperfectly, to do so. Wingtip mounts became common on many jet fighters as it allowed ready missile mounting, did not displace underwing stores, and somewhat reduced overall aircraft drag. (Artwork by Tom Cooper)

This Egyptian MiG-21MF carries four R-3S rounds and a centerline fuel tank for a combat air patrol mission. The MiG-21 was not initially designed for AAM. The MiG-21F-13 model was specifically developed to carry two of the R-3S missiles, but retained a gun (albeit with few rounds). Subsequent models deleted the gun, but this was soon demonstrated in combat to have been a non-optimal decision. The missile, an inferior copy of the American AIM-9B Sidewinder, was so unreliable that the ability to carry four rounds increased the chances for hitting a target. The MiG-21MF also got its cannon back. (Artwork by Tom Cooper)

The United Kingdom's first successful AAM was the IR-guided Firestreak with is unusual faceted nose transparency. It was a respectable product but saw service with a few types produced and exported by Britain and so only 3,000 to 4,000 rounds were manufactured. Of these were advanced models of the Gloster Javelin, an example of which is this Royal Air Force Javelin FAW 9 seen here with Firestreak rounds on outboard wing pylon/launchers. (Artwork by Tom Cooper)

This Ecuadorian Jaguar models an uncommon over-wing AAM installation, mounting French Magic missiles. While avoiding interfering with heavier underwing stores, mounting a missile atop the wing was cumbersome and time-consuming. The third-generation Magic was a reasonably capable AAM, but its deficiencies were addressed with the improved Magic 2. Both were carried on several models of fighters, typically with wingtip and underwing mounting, including by other operators of the Jaguar. (Artwork by Tom Cooper)

The Israel Defense Forces/Air Force (IDF/AF) had to stretch its funding and so relied principally on cannon for air-to-air combat. Yet, attempting to stay abreast of current aviation technology and so explored the AAMs offered by France, the only country willing to sell them warplanes. Israel also developed its own IRH missile in the first-generation Shafrir. This IDF/AF Mirage IIICJ is armed with the initial model Shafrir under the wingtips and a French second-generation radar homing R.530 under the belly. The Shafrir proved to be a great disappointment and only around 150 were assembled. The radar homing R.530, that also proved largely ineffective, was to work with the Mirage's radar that was itself of low reliability. The missile was heavy and reduced aircraft maneuverability to a discouraging extent. After a spate of combat disappointments it, too, was retired. (Artwork by Tom Cooper)

An Israeli Vautour IIB, employed in reconnaissance overflights of enemy territory and air defenses, is armed with an indigenous Shafrir AAM for self-defense. In similar installations elsewhere, for bombers and patrol aircraft, this offered the aircrew a means to 'fight back' if attacked by interceptors before friendly fighters could respond. However, the likelihood that this would be effective against a far more maneuverable interceptor, unlikely to present itself for an engagement, makes it more of a 'morale booster' measure. This was especially true of the Israeli aircrew given the exceptionally poor performance record of the first-generation Shafrir. The decades-old ambition to fit AAM to bombers, especially in the USA, was never satisfied. (Artwork by Tom Cooper)

This Libyan MiG-23MLD carries a large R-24T under the wing and R-60MKs under the inlet. The aircraft would commonly be armed with both radar homing R-24R and IR-homing R-24T rounds to permit tactical flexibility depending on engagement circumstances. Developing dual-purpose missiles of the same basic airframe was a common Soviet practice for decades. The R-60MK was a 'dogfighting missile' of much less mass and greater maneuverability than the R-24. There was very little ground clearance between the bottom R-60 round during landing and taxi, making it vulnerable to damage. Such installations, in close proximity to inlets, risked smoke ingestion by the engines which could promote compressor stall and flameout – a grave event during combat. (Artwork by Tom Cooper)

South Africa had difficulty acquiring foreign-built weapons owing to its racist policy of apartheid. It worked to develop its own armaments that included AAM. As with many nations, its first attempts focused on IRH and leveraged off the Sidewinder example. The earliest successful product was the second-generation V3B Kukri that was adopted by the Mirage F.1 squadrons. In combat over Angola the South African Air Force fired several V3Bs (shown here on a Mirage F.1CZ), but with discouraging results. Approximately 900 rounds were manufactured. (Artwork by Tom Cooper)

A Saudi Arabian Tornado ADV is equipped for combat air patrol with four very capable IR-guided AIM-9Ls under the wings and four radar-guided Sky Flash AAM under the fuselage. This mix of two missile guidance types became typical, once engine power permitted such weight, for tactical flexibility. The fourth-generation AIM-9L was the first true all-aspect IRH system. The Sky Flash was the first inverse monopulse SARH system. The fourth-generation Sky Flash had to fit semi-submerged mounting cavities initially designed for the American AIM-7 Sparrow AAM, and so the missiles look very similar. This sort of design constraint is not uncommon. (Artwork by Tom Cooper)

The F-14 Tomcat was one of the few aircraft that could accommodate three AAM types. This US Navy F-14A carries two AIM-9Ms and two AIM-7Ms under the wings and two AIM-54C radar-guided missiles under the fuselage – all fourth-generation AAM. The AIM-54 Phoenix was a missile of impressive capabilities that included active radar homing, datalinks, and lofted flight path for exceptionally long range. The aircraft systems to operate all of these missiles, plus a cannon, was very complex. Operated by the Iranians during its war with Iraq, the Tomcat was proven an invaluable weapon. This aircraft also has a Television Camera System senor package beneath the nose; a TV telescope to assist in long-range identification of targets. (Artwork by Tom Cooper)

The Iraqis fielded a wide assortment USSR combat aircraft during the 1960s and 1970s, some armed with Soviet AAM, and these were felt to have significant deficiencies. Early in its war with Iran during the 1980s Iraq acquired French warplanes. Among these were Mirage F.1EQs, equipped with the initial model of the IR-guided Magic (third generation) and the radar homing Super 530F (also third generation). These proved more effective than the Soviet AAM, but the Iraqi fighter pilots fought so cautiously that they could not exploit the best advantages the Western missile had to offer. However, the Super 530F, when employed with special tactics, shot down three of the invaluable Iranian F-14As in the span of nine days. This is an example of a brief tactical advantage gained from an advanced AAM introduced into a combat milieu. (Artwork by Tom Cooper)

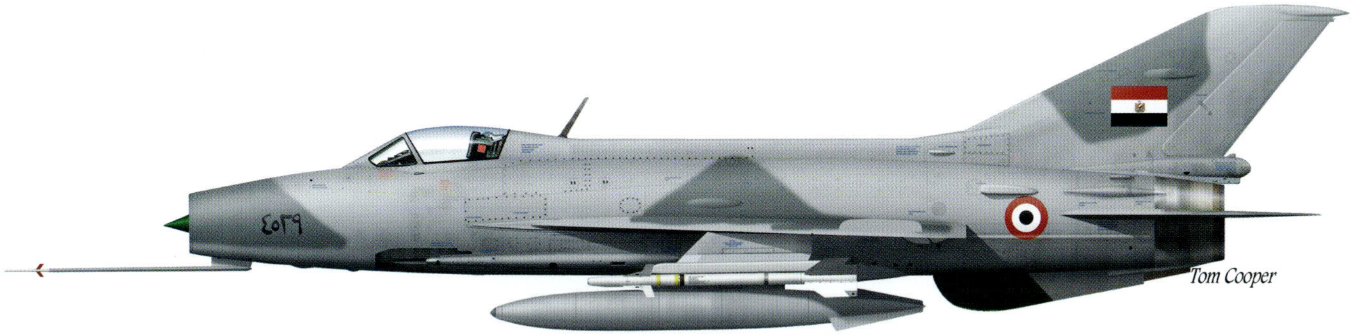

Air-to-air missile development generally kept pace with fighter aircraft development. In some cases new aircraft types had to accommodate older missiles and in other cases the missile was designed to fit existing platforms. Particularly older jets, more costly to replace than AAM, benefitted from adopting a more modern weapon. Occasionally these saw AAM from other nations fitted to a foreign mount. An example is this Egyptian F-7B, a Chinese knock-off of the Soviet MiG-21, carrying an American third-generation AIM-9N Sidewinder. (Artwork by Tom Cooper)

Some fighters enjoyed a long service life and adopted several AAM models during that period. An example of this is the Soviet Su-15 that entered service in the mid-1960s armed with R-8 missiles of IRH and SARH variants. As the Sukhoi interceptor was upgraded so were the AAM, first as the R-98 and the R-98M. This sustained the platform into the 1990s, though by then obsolescent. This Su-15TM is as it appeared in 1983 armed, on the port side, with an IR-guided R-98MT and 'dogfighting' R-60 inboard, plus a 23mm gun pod on centerline. (Artwork by Tom Cooper)

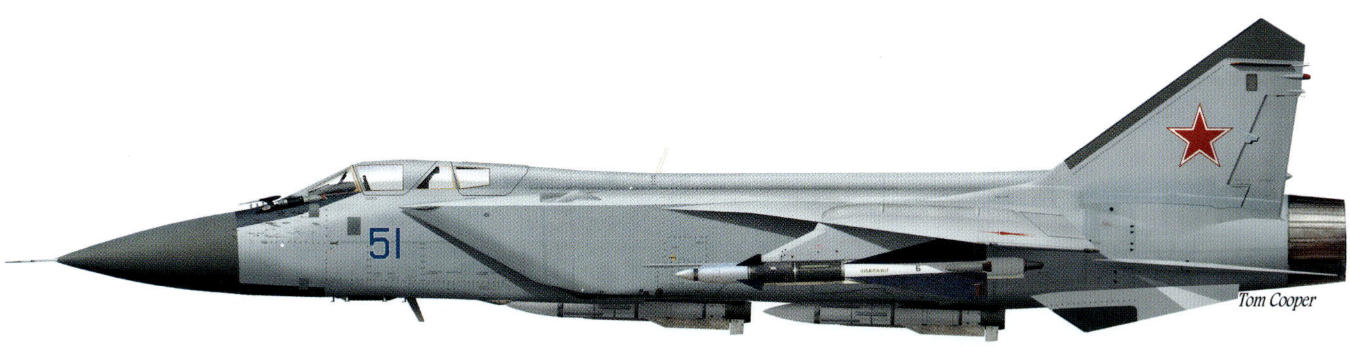

Very long-range AAM began appearing with third-generation weapons during approximately the 1970s. These tended to be large weapons, the Soviet R-33 being one example at approximately 1,080lb (490kg) launch weight and 14ft (4.3m) length. Four of these semi-active radar homing weapons were carried semi-submerged beneath the equally large MiG-31 interceptor. The aircraft shown also has third-generation, medium-range, radar homing R-40RDs that were also big at approximately 1,000kb (450kg) and 22ft (6.8m) length. While not exceptionally maneuverable, the MiG-31 was very fast and, armed with long-range AAM, it was a formidable threat. (Artwork by Tom Cooper)

This Saudi EF-2000 carries radar homing AIM-120 AMRAAMs on the fuselage stations and IR-homing IRIS-Ts under the wings. Compared with the prior Tornado and its missiles, this combination represents the next generation of both. The fourth-generation AIM-120 was an active radar homing round and the German IRIS-T a fifth-generation IR imaging system of high agility. The aircraft is also equipped with an infrared search and track (IRST) system with sensor just ahead of the windscreen on the port side of the nose. This permitted longer range visual identification of a target before firing. (Artwork by Tom Cooper)

Although Israel developed AAM of its own, it continued to buy American products to gain performance advantages and in case some products were briefly 'grounded' owing to uncovered deficiencies. Among the IRH weapons were the Rafael fourth-generation Python 4 matched with an American radar-guided AIM-120A. These are shown on an IAF/AF F-16I in a common configuration of the AIM-120 on the wingtip launchers and Pythons inboard. This arrangement saw much use during the 2023 though 2025 war when Israel was beset with hundreds of inbound drones, reflecting a growing demand for 'drone busting' also reflected over Yemen and Ukraine. The remarkable aspect of these engagements is destroying a tiny aircraft costing tens of thousands of dollars with a missiles costing half a million to a million. (Artwork by Tom Cooper)

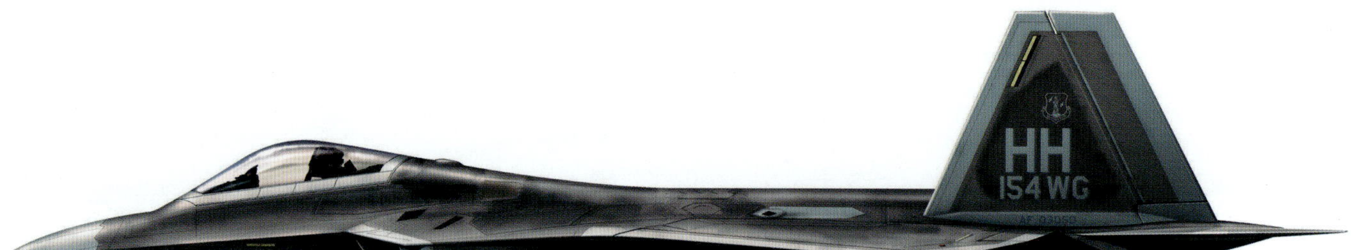

Internal carriage of stores, including AAM, reduces drag and enhances low-observable (stealth) properties of the fighter. Although undertaken early in the age of AAM in the USA, it disappeared for a time until embraced anew with the F-22A. This aircraft carried AIM-9 IRH missiles in side weapons bays and AIM-120 ARH rounds in belly bays. The Sidewinders were extended from open side bays on a trapeze launcher, angled outboard. (Artwork by Rolando Ugolini)

The stealthy F-35 fighter-bomber also had weapons bays meant principally for air-to-ground ordnance. When carrying AAM these were also accommodated in the belly bays. Here a South Korean F-35A is deploying a fifth generation AIM-120C via jettison. The AIM-120C variant of the Advanced Medium Range Air-to-Air Missile (AMRAAM) had reduced-span aero surfaces supporting internal 'compressed' carriage. This principle was followed by other teams fitting AAM to weapons bays. (Artwork by Tom Cooper)

China's effort to develop competitive AAM proved a multi-generational program as the nation built industrial capacity, climbed a technology learning curve, and trained engineers suitably. They began by manufacturing and then improving Soviet weapons followed by Western products. In some cases these projects were unauthorized and supported by industrial espionage. Aircraft systems also had to mature to fully exploit the most modern missiles. In the 2000s this long effort began to give birth to world-class AAM of all varieties. This J-10C, developed from the Israeli Lavi fighter, is armed with radar homing PL-15 missiles inboard and IR-homing PL-10s outboard. (Artwork by Tom Cooper)

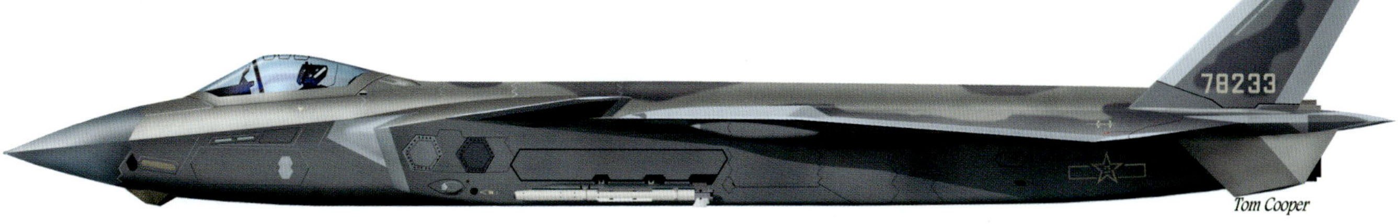

As China began developing stealthy aircraft it had to accommodate AAM inside weapons bays. This J-20 has deployed a PL-10 from a side bay. Unlike the older American F-22A, the J-20 design allows the bay doors to be closed while the AAM is extend on its trapeze launcher, restoring some of the original low-observable properties. Other existing AAM were likely to require compressed carriage models to fit these newer jets. (Artwork by Tom Cooper)

Arming helicopters with AAM helped to discourage intercept attempts by enemy fighters while also giving aircrew a gratifying means to defend themselves. Engaging enemy helicopters was another potential use. However, given the low payload capacity of helicopters, the addition of an AAM represented a significant performance impact along with added complexity and subsystems load. They displaced primary mission payload and could encourage aircrew to 'go looking for trouble.' The weapons selected tended to be derived from man-portable air-defense systems and so were very small and of short range. This US Army AH-64 Apache attack helicopter has a two-round Air-to-Air Stinger (ATAS) installation attached to the tip of the port stub wing, the system coming from the FIM-92 Stinger shoulder-launched weapon. The capability was seldom carried given the plentiful air cover the US Army could expect. (Artwork by Pablo Albornoz)

In some cases adapting AAM to helicopters was a combat expediency that could as often be successful as unsuccessful. The quickest option was to employ common fighter aircraft AAM, though these were so large as to generally negate other weapons on the helicopter. The Americans was able to mount Sidewinders onto Marine Corps AH-1T attack helicopters during confrontations with Iranians in the Persian Gulf during the 1980s. Though successful, the opportunity to fire the Sidewinder never presented itself. The Iranians had attempted a similar arrangement, mounting an AIM-9E on an Iranian Army AH-1J. In this case the goal was to shoot at other helicopters. The project had to be abandoned when it became clear the Air Force was unwilling to give up any of its precious Sidewinders. (Artwork by Tom Cooper)

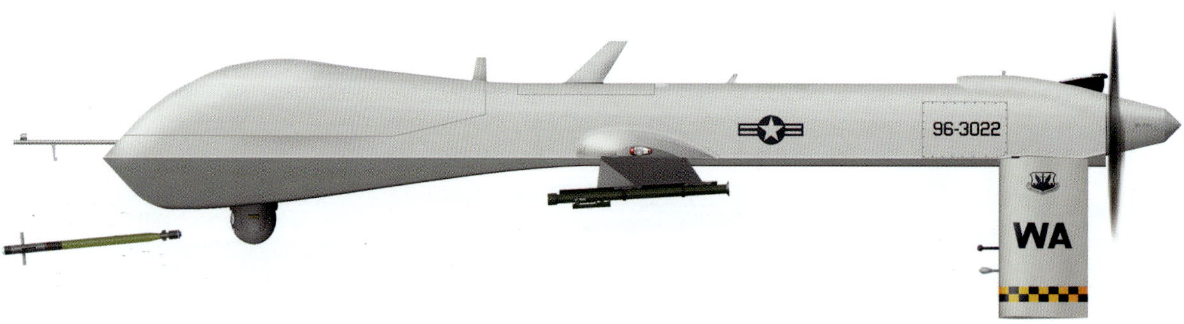

The advent of armed uninhabited air vehicles (UAVs) naturally led to thoughts of adding AAM to the weapons mix. This did not catch on immediately owing to the low payload capacity of the early craft and their low cost, being essentially expendable. However this changed in time as broadened missions were envisioned. This US Air Force RQ-1 Predator drone is armed with an Air-to-Air Stinger (ATAS) launcher on the weapons pylon for a special mission. More advanced UAVs were expected to carry such weapons internally. (Artwork by Tom Cooper)

By the time such weapons became available the most likely nuclear bombing tactic was to send aircraft individually on separate tracks to their targets distributed across a vast nation, a formation or stream of bombers being rendered anachronistic by the weapons they were delivering. None-the-less, the nuclear-armed AAM remained. Only the USA fielded such weapons (two), other nations only contemplating same.

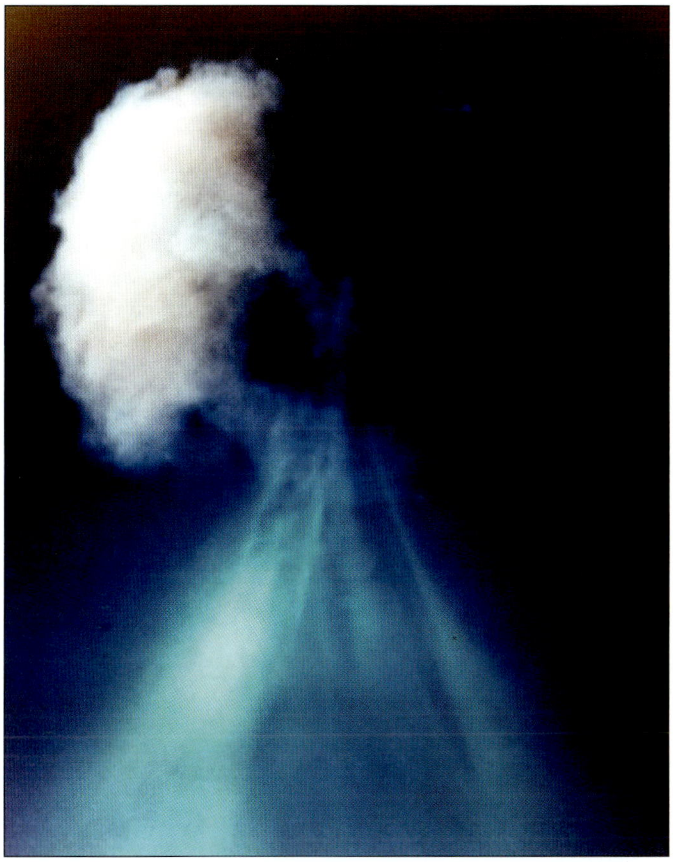

The result of the 2-kiloton air burst test of a Genie air-to-air weapon (Volume 3, Chapter 1) at 19,000ft (5,800m) altitude in 1957 is shown. The lopsided white cloud is more than 0.5mi in diameter and the effects of the blast extended far beyond this. (USAF)

There are AAM without warheads with expectation of a direct hit on a target (hit-to-kill, HTK). Eliminating the warhead can, of course, shave a significant mass off the missile – as much as 10 percent or more – but demands considerable accuracy. Other than first-generation designs expecting a hit to set off a warhead of modest size, the HTK technique was not adopted until fifth and sixth generation systems. It had quickly become clear with the first-generation systems that a direct hit was a remote probability with then-applied technology, and the airspeed of 2 Mach or less would mean an impact of kinetic energy insufficient to have high probability of a kill. The later generation AAM flew considerably faster and had much higher accuracy, making HTK a more reasonable expectation. By then HTK had been adopted by some SAM systems, however translating the same capability into the confined volume and power available for an aircraft and missile was nontrivial. It is still a rarely considered technique.

Seeker Head

The seeker head is the missile section that contains the device detecting and tracking the target aircraft via various means. Several parts of the electromagnetic spectrum (Figure 34) are more useful for detection purposes than others given inherent energy and detection methods. These propagate from their sources as wave phenomena, at different wavelengths (Figure 29), and at the speed of light – 142,786mi/sec or 299,792km/sec, essentially instantaneous for the purposes of AAM. Their practicality as a means of missile seeker homing is a function of the intensity of their emissions from a target towards the sensor compared with background sources, that is signal-to-noise ratio (S/N). The S/N must be greater than 1, usually much greater, for effective detection and tracking. The radiated signal attenuates over distance owing to scattering in the atmosphere, with some wavelengths affected more than others. Seeker sensitivity and signal processing is critical in distinguishing the target.

Seeker data are employed by a guidance system to bring the warhead within lethal radius of the target. They break down basically into optically guided and radar-guided varieties. Homing can be passive (no EM emissions from the AAM) or active (with emissions). The ability to resolve closely spaced targets in a

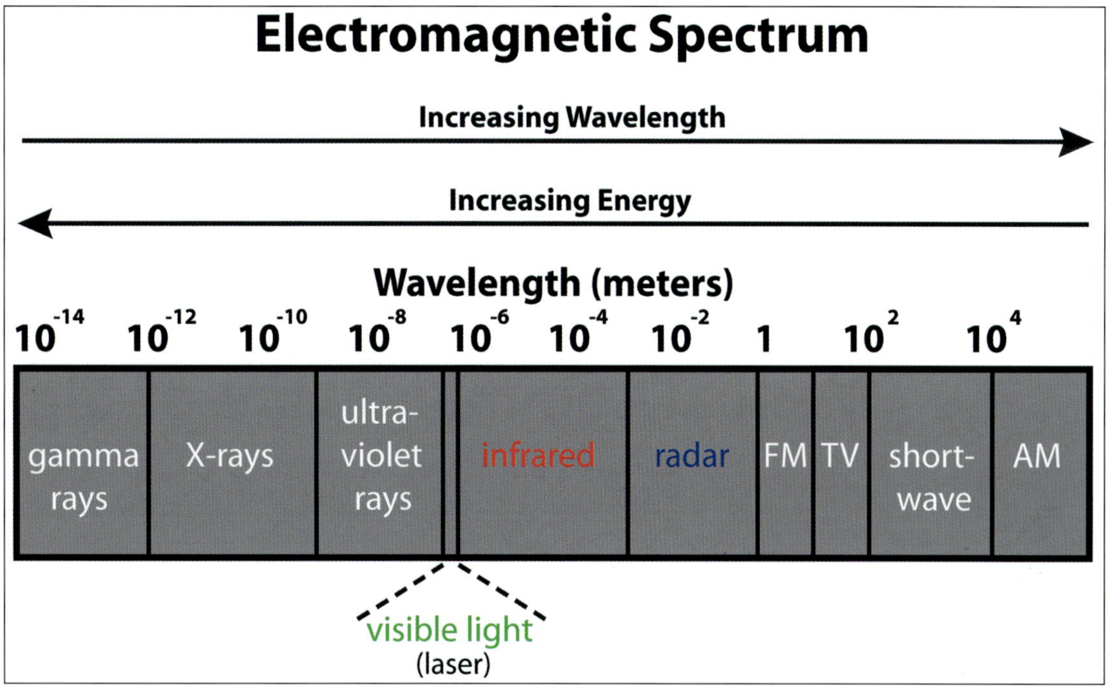

Figure 34. The principal segments of the electromagnetic spectrum are shown. People only detect visible light (also used for lasers) whereas radar and infrared wavelengths are commonly employed for AAM guidance. (Drawing by author)

formation of aircraft to attack a specific target can be a demanding requirement of the seeker. The seeker may switch between targets depending on which presents the greatest return via glints, with the resulting jerky flightpath draining energy and decreasing range. In the endgame it might be commanded to a compromise between the strong reflections and miss all, or guide to the one with the most ideal final signature. Each detection and tracking means employed or conceived will be presented.

Aircraft radar is often essential in the initial detection of an airborne target. It may then be used as one means of guiding an AAM. Pulsed radar signals can be used to search for targets, the antenna sweeping the sky, giving basic bearing and range (time between the transmission of the pulse and reception of the reflected signal). The FCS can then communicate to the missile seeker where to find the target. Difficulties arise with varying return signal strength given target maneuvering (altering RCS) and the radar may be fairly easy to jam. Measuring Doppler Shift (Figure 28) is fundamental to radar target detection against non-moving or slow-moving background. This employs continuous wave emissions as opposed to the pulsed wave. The CW then 'locks-up' the target, the antenna tracking the target, to calculate Doppler characteristics like velocity. Locking-up the target can also alert the targeted aircrew that a launch is imminent or has occurred provided they have a suitable radar warning receiver (RWR, also known as Radar Homing and Warning System, RHAWS). Pulses at different repetition frequencies (PRF) can improve range and Doppler determination, and better discriminates between various targets. This pulse-Doppler (PD) technique eliminates the need for also having CW to lock-up the target.[12] The next innovation, the monopulse-Doppler, is better able to calculate bearing and range from a single pulse against clutter and is less susceptible to jamming.[13] With the shooter having locked their radar onto the target, tracking of additional threats is negated and some situational awareness is lost. This was resolved in the 1970s with introduction of the track-while-scan radar capability. Hence development and maturation of radar and FCS technologies were a parallel effort with AAM history. But, this and radar operation lies beyond the scope of this text.

Initially the operation of onboard radar and interpreting the scope images, designating a target, operating the FCS, and ensuring the AAM was ready to fire were of such high workload that a second aircrew member (weapon system operator, WSO or 'wizzo', or radar intercept officer, RIO or 're-oh') was necessary in a two-seat fighter aircraft. It was several decades before technology and automation permitted the workload to be reduced to the point where single aircrew operators were practical and effective.

Command or Preset Guidance

As the title implies, this is a missile fired with a course pre-programmed into the guidance system or uploaded just prior to launch. Against a moving target aloft like an aircraft this is ineffective. However, for long-range missiles this can be used to direct the AAM towards the target, usually detected by aircraft radar, until its shorter-range homing system acquires the target and begins providing signals to the guidance/navigation system. Transmitted radio datalinks from the aircraft FCS can send mid-course flightpath corrections analogous to command guidance. A command link for the entire flight is possible but impractical given inaccuracies at the most distant endgame engagement. The datalink can limit the number of targets that can be engaged simultaneously unless the system permits transmissions coded to specified missiles and transmitted sequentially.

Manual Command Line-of-Sight (MCLOS)

This was among the earliest systems employed because it could leverage off the technology developed during the Second World War to guide freefall bombs with 'precision.' In this case the sensor is the pilot's or gunner's eyeballs, and guidance is displacement of a cockpit control altering a radio signal to the missile (remote control) to alter the flightpath to match perceived closure on the target (Figure 35). The control is typically a joystick commanding pitch and yaw of the missile in flight. An alternative is a very long and thin wire, or wires reeling out from the missile and connected to the launch aircraft to carry the command signal. The round typically has a flare at the aft end that the operator uses to guide the AAM to the target once the motor with is attendant smoke burns out. It was a 'simple' matter of keeping that glow on the target or on a perceived closure path.

The greatest deficiency in this system is that the operator has great difficulty judging distance between the missile and target, especially as the former appears so tiny at a distance. In the effort to bring the missile in line with the target it could well be ahead of the aircraft and so overshoot. Additionally, maintaining LOS means daylight engagements only and minimal atmospheric obscuration like haze or clouds. Even smoke left behind by the missile confounds the task. The system has typically relied on a direct hit rather than a PF. Flight testing has shown MCLOS effectiveness to be very low. Regardless, such systems were fielded briefly in the early post-war years because no others were available.

Electro-optical (EO) technology permits television (TV) or other imaging to be projected from a missile in flight to a display in a cockpit where an operator can steer the missile onto the visually identified target. This has been used in air-to-ground ordnance. While it allows manual steering to a visually tracked target, an air-to-air application is ineffective. The latency inherent in the transmission, display, observation, reaction/control input, missile flightpath correction (even assuming very fast computer processing) makes manual guidance of a missile at a supersonic closure rate on a small and maneuvering target an impractical task.

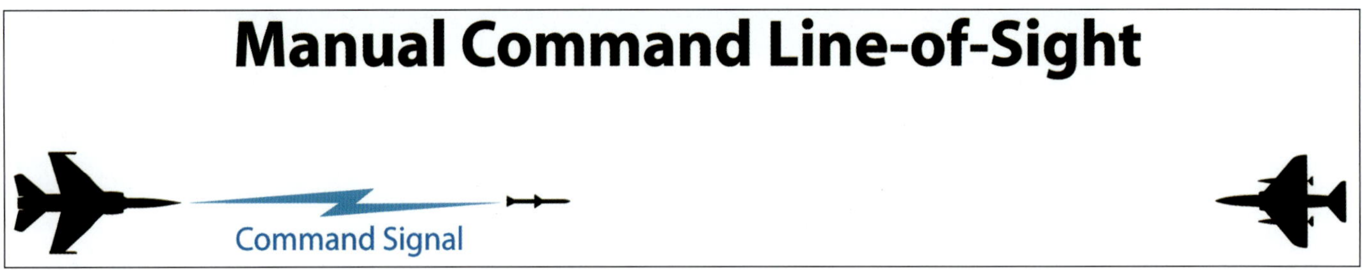

Figure 35: A depiction of the basic function of a manual command line-of-sight (MCLOS) air-to-air missile systems shows the signal from the attacking aircraft to the AAM, usually via RF emissions, guiding the missile to the target. (Drawing by author)

Passive RF Homing

Passive homing on radio frequency emissions from the target aircraft (commonly radio or radar) has been pursued (Figure 36). This has been referred to as an anti-radiation missile (ARM) and although commonly employed air-to-ground to target SAM radar, there have been a few examples of air-to-air ARM. Especially the emissions from fighter fire control radar can make them vulnerable to such weapons. Destroying high-value and large airborne aircraft such as airborne early warning (AEW) and Airborne Warning and Control System (AWACS) and Airborne Early Warning and Control (AEW&C), or dedicated jamming platforms, which are usually fielded in low numbers but are high-intensity emitters, could markedly contribute to combat success. Guiding on the jamming signal from a platform seeking to counter a radar-guided AAM, and so a like signal, is termed home-on-jamming (HOJ) or 'jam track.' This mode has been featured in many radar-guided missiles as a secondary guidance means in an electronic countermeasures (ECM) denial environment should the jamming successfully confound the principal guidance.

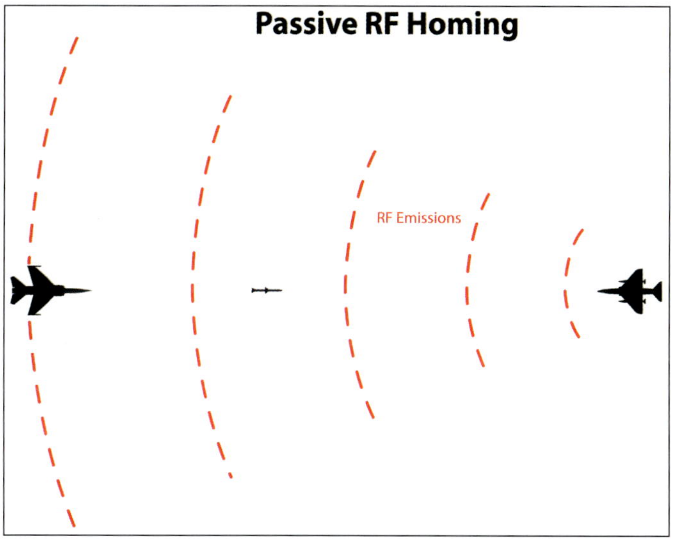

Figure 36: A depiction of the basic function of a passive RF homing AAM shows the original RF emission from the target aircraft on which the missile guides. (Drawing by author)

Accuracy of such weapons are greater than semi-passive SARH types (see later) because the signal to which they are homing grows in strength the closer the AAM approaches. This may extend range closer to the full kinetic capability of the missile. Not being reliant on radar returns, this has the potential for longer range engagements at some aspect angles should the AAM have such reserve performance. With HOJ, the jamming is most likely tuned to the characteristics of the radar being used as guidance. However, specialized aircraft like AWACS transmit on a large number of frequencies, and not necessarily continuously, making the choice for guidance and discrimination potentially challenging. The AAM has to be 'tuned' to the specific frequency or a few select frequencies known to be emitted by the target, or even finer characteristics like modulation and pulse repetition frequency. In such cases discontinuing or altering the transmission will cause the ARM to cease guiding and 'go stupid.' For HOJ, guidance should return to SARH.

With no emissions of its own, these weapons are 'launch-and-leave' or 'fire-and-forget' – though few fighter pilots would 'forget' to look and see if their missile struck the target. Apart from being alerted that opposition aircraft are in the area, the targeted aircrew might be unaware they are under fire except possibly seeing the rocket plume during the initial portion of the flight. By the twenty-first century large aircraft had the means to detect and decoy an inbound missile (Chapter 4), but fighters remained reliant on visual detection before activating decoy means. Decoying may be practical by providing an alternative and more powerful emission source, provided high confidence they know the channels the AAM is targeting. This may be done with a trailing decoy (following on a wire) with emission source. The large aircraft are less likely to evade via maneuvering than a fighter. The more practical protection for the large, high-value targets were their detection of the inbound fighter and then vectoring of counter-air assets to engage attackers before they come within range to fire the AAM. Ceasing to emit if warned could save the aircraft.

Other Passive EM Homing

With many elements of the EM spectrum to choose from (Figure 34) it is possible to employ other autonomous passive tracking means.

Ultraviolet light (UV) permeates the atmosphere and an aircraft, when seen against this background, is a UV 'hole.' This contrast can be used for tracking, but, in AAM applications, some IRH seekers have used a narrow UV band as a second 'color' for target tracking as another means of discriminating countermeasures. The common UV detector uses cadmium-sulfide (CdS) for the 0.3–0.4 micron bandwidth (micrometers or a millionth of a meter, 10^{-6}, μm) and operates with the same optical techniques as with IRH (see later).

The visible light contrast tracking technique has been successfully employed in air-to-ground precision-guided munitions and in the Soviet Strela-1 surface-to-air system. The latter system operated in the 0.4–0.7 microns bandwidth and was only useful with the aircraft against a clear sky background. While it allowed an all-aspect capability at a time the IRH systems could not (see later), the limitation was too great. Even constrained to an aft aspect with a minimal capability, IRH system was assessed as more effective.

More recently EO contrast tracking has been used in air-to-air applications such as an uninhabited air vehicles (UAV) closing on an airborne tanker – a low-speed scenario. With sufficiently high resolution it is possible to identify an aircraft and aspect angle by matching the detected image to certain features and characteristics of stored data. This requires considerable onboard digital processing and at high-speed to permit this to serve for tracking and proximity fuzing at missile engagement velocities and against maneuvering targets. It could be limited to good lighting conditions for visual light or require multiple spectra to include IR. This has been employed in some fifth generation imaging-IR missiles (see later) to identify the most vulnerable spot on a target to guide to and fuze. To the author's knowledge, no such system is known to have been used as the principal guidance means for an AAM.

Particularly the variable and far slower propagation of sound waves (not EM), especially at supersonics airspeeds or at low level, make this choice problematic. There is less reverberation in flight than on the ground, but at low-level the sound waves reflecting off the surface could still confuse such a seeker. The AAM is likely to end up in a lag-pursuit intercept (Figure 24), overshooting the intercept behind the target and then entering an inefficient tail chase.

Beam Rider (BR)

The beam riding technique was employed in early missiles during the period when electronic miniaturization remained in its infancy and it was very challenging to employ more accurate radar homing techniques that relied on energy reflected off the target. It leveraged off the automatic tracking radar for directing gunfire perfected during the Second World War. Beam riding classically employs radar, either aboard the launch aircraft, on another aircraft, or a ground station, to illuminate the target. The missile homes on this outbound beam from the source (Figure 37), not radar energy reflected by the target (also called back scatter). Consequently, it is essential that the tracking signal be locked onto the target either manually or automatically for the entirety of the missile fly-out. The missile commonly has several rearward facing receive antennas to permit an electronic comparison of receive signal strength. The difference ('error') between these measurements is used in the guidance algorithm to steer the AAM such as to drive the error to zero and so remain in the center of the beam. In some BR applications these antennas have been at the extremities of the aero surfaces. For a four-fin missile, the antennas are paired on opposite sides to provide data for two-axis maneuvering. Three antennas just require different data processing. Other applications have had antennas embedded on the sides of the missile body or at the rear. The missile must have some manner of PF to initiate warhead detonation at closest approach to the enemy aircraft. Suitable accuracy for an impact detonation has been elusive for this technology.

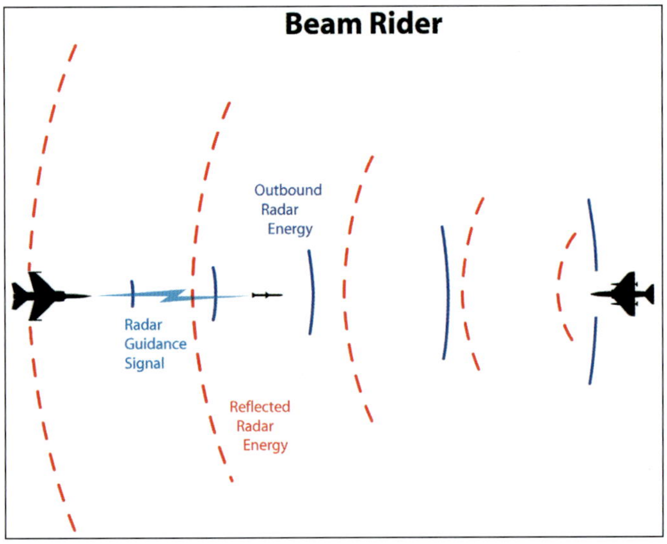

Figure 37: A depiction of the basic function of a beam riding AAM shows the original radar emission from the attacking aircraft and reflected back from the target, but the missile guiding on the original outbound emission. (Drawing by author)

The missile must be launched such that it enters the transmitted radar beam (it is 'gathered'), narrow near the source, to begin tracking. It is possible to fire a volley of BR missiles into the beam to increase probability of kill. A surveillance mode of the radar system may be used initially to ensure a wide beam (a few degrees) for this initial phase of flight before narrowing to the tracking beam mode (as little as 0.1deg). However, the beam widens and signal strength diminishes as range increases from the transmit antenna (inverse square law, Intensity $\approx$ 1/Distance^2). The widening beam also means that while the AAM may remain within the beam it may not approach the target close enough to trip the proximity fuze. The pilot must also ensure that any aircraft maneuvering or radar antenna steering to keep the beam on the target is done gently such that the AAM can follow without 'spilling' out of the beam and breaking lock. The choice of radar operating bandwidth effects power available, operating characteristics, and effectiveness in weather. Additionally, radar ground clutter can confound the effectiveness of the system, making low altitude engagements generally impractical. Consequently, S/N plays a role in the range and accuracy of the beam rider. All this makes tracking the target with suitable accuracy for an actual hit problematic beyond about 2.5–3.0mi (4–5km), and against a maneuvering target exceedingly remote. A large bomber might be a suitable opponent. A separate onboard target tracking means for endgame engagement lends higher accuracy and so greater chance of coming within lethal radius for warhead effect. This can be a low-power radar or IR seeker. However, there have been no production examples of such a combined system to the author's knowledge.

Another limitation is that the beam riding AAM is in a constant chase (curve-of-pursuit) and cannot execute lead collision engagements (Figure 24). This extends required engagement time and so energy dissipation, potentially shortening range/flight duration against a maneuvering target. It usually also means the final intercept maneuvers are very tight and possibly beyond the remaining energy capacity of the AAM. This can be overcome with a radar tracking the target and another guiding the missile for the lead-collision course, with that navigation solution processing performed at the emitter platform. This has generally been impractical for a fluid air combat scenario.

More recently a laser beam has been used with this technique, either as a semi-active technique (as in air-to-ground munitions) or active technique, the guidance seeking to keep the AAM centered in the laser beam or reflection. With some aircraft equipped with ranging and targeting laser systems, an AAM interoperable with those systems would be beneficial. Another benefit is freedom from the common countermeasures deployed by aircraft. A laser beam generated from the missile itself is generally not practical given that the target must be initially detected by some other means that could, itself, serve for homing. Generating the energy for a suitably powerful laser is especially difficult onboard the missile given confined volume. Even when emitted from the launch platform or some other source, holding the beam steady on the target without excessive 'jitter' over miles of distance, and the target potentially moving at a high cross-path rate, is exceptionally challenging. Laser light is also scattered in the atmosphere across even moderate distances, inbound to the target and then in the return, and so limiting the missile to short ranges. Consequently, the few examples of such systems have been small and short-range missiles launched from helicopters and UAVs (Volume 5, Chapter 2).

Semi-Active Radar Homing (SARH)

Radar has been the principal guidance type for BVR and all-aspect engagement given that radar returns are practical at such distances with sufficient transmit power. Radar is less affected by weather compared with IRH and so is superior for all-weather weapons. The SARH missile has only a radar receiver, unlike the ARH system (see later), and so relies on the launch aircraft radar to illuminate the target that reflects the energy to the AAM. Depending on the capabilities of the radar and the kinetic performance of the missile, these have typically been medium-range systems with a maximum normally defined as 15mi (24km).

For SARH, the missile steers to radar energy reflected off the target, originating from the launch aircraft or conceivably another source (Figure 38). This was most commonly a pulse-modulated radar signal to locate the target and then a continuous wave (later monopulse-Doppler) to lock it up and provide radar reflections for the AAM to steer to. The seeker receive antenna (forward 'echo' antenna), rotating in a scan pattern, is usually uncaged prior to launch and the antenna 'slaved' to aircraft trackers to 'cue' the target (see earlier) and reduce acquisition time at launch. This has typically been performed with a cockpit controller to designate a 'blip' on the radar display, the FCS then steering the seeker to that elevation and azimuth. The missile indicates that it has locked-up the target. During the fly-out the target relative position (strongest radar return) is obtained by AAM antenna scanning. Whenever an error exists between the antenna boresight and the target an internal signal modulation is generated. This signal is processed to obtain a tracking command for the front antenna and a guidance command for the flight control section.

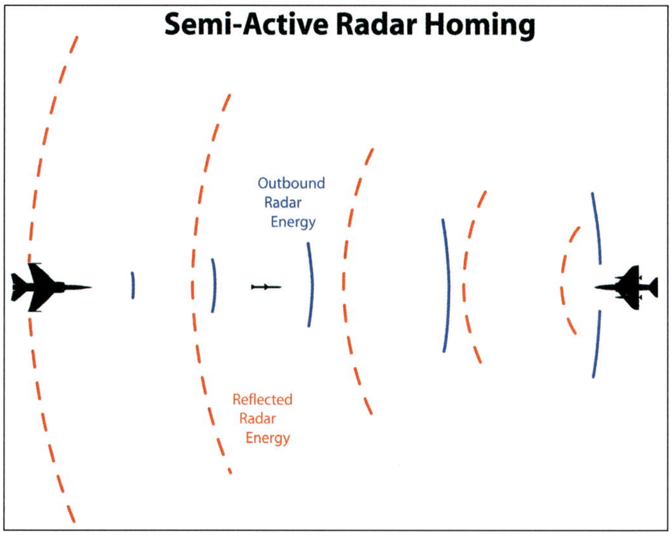

Figure 38: A depiction of the basic function of a semi-active radar homing AAM shows the original radar emission from the attacking aircraft and reflected back from the target on which the missile guides. (Drawing by author)

Aircraft radar are designed with a number of search modes to locate targets based on the local environment and tactical scenarios. Once a target is located the radar can be commanded to track it with continuous illumination rather than periodically in the search mode. The AAM receiver is tuned to the radar emissions by the FCS prior to launch. The missile will commonly be required to lock onto the reflection of that signal prior to launch. The first second or so of flight may be 'on boresight', rejecting guidance commands, until the missile is beyond the influence of the aircraft radar sidelobes that may generate spurious signal. There have also been designs with the missile fired without lock, some with datalink flight updates, to fly out until receiving the radar signal reflected from the target as the range shortens. The reflected energy from the target consists of the RF carrier signal, the target/AAM Doppler (range and velocity), and the launch aircraft/target Doppler. Usually, the AAM is also receiving the reference radar signal from the launch aircraft via an aft (direct) antenna, giving the reference and the launch aircraft/AAM Doppler (Figure 39). Comparing the reference and the Doppler from the target gives range and closing velocity, the latter used to adjust the proportional navigation g commands (K factor, Figure 27). This reference is used in processing the received radar reflection to isolate the AAM/target Doppler for greater guidance accuracy. The target Doppler is proportional to the closing velocity with frequency higher or lower than the RF signal depending on aspect, or the same if directly abeam. One approach is for the AAM to compare the frequency modulated (FM) reference signal from the aft antenna with the time-delayed ranging signal reflected by the target, the frequency difference being proportional to target range determined via Doppler difference. This enables the tracking. Range

The aircraft radar, especially its power and transmit modes, are crucial to success with a SARH missile. Here the radar antenna of an F-4 Phantom is exposed, the mechanic installing armor between the unit and the aircraft forward bulkhead. (USAF)

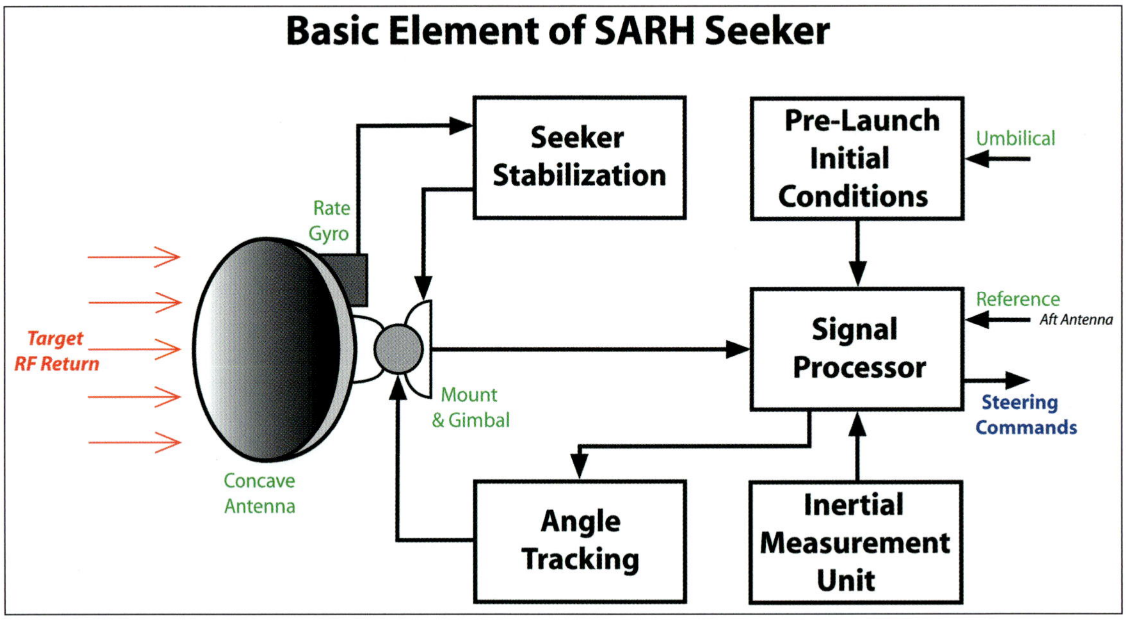

Figure 39: The basic elements of an early SARH missile guidance section is shown. More advanced systems might include a datalink receiver and a digital processor, the latter especially if the AAM has a pulse-Doppler receiver. The pre-launch initial conditions might include target range and Doppler. (Drawing by author)

can also be uploaded from the aircraft's FCS prior to launch (Table 4) and then updated during the AAM's flight via Doppler calculations that can also resolve speed (the Speed Gate) to discriminate between multiple targets that may include countermeasures and returns off the ground ('clutter'). This also defines a narrow band of reflected frequencies in scanning for the target, the Doppler Gate.

Since the SARH seeker is receiving energy first traveling to the target and then reflected back, the intensity is even less than for a beam rider and possibly passive seekers. Instead of the inverse square of the beam rider it is to the inverse forth power (Intensity ≈ 1/Distance^4). The reflected energy is greater as the AAM speeds toward the target and is greatly dependent upon the power of the aircraft emitter. It may be too weak initially and so some missiles fly a set heading initially until picking up a signal of sufficient strength. The range and the target's radar reflectivity can determine the strength of the return signal. This can be further degraded by how the radar energy is reflected as a function of aspect angle to the incident radar waves. Some portions of the aircraft or materials will reflect directly back or, by coincidence, towards the missile receiver while other portions will reflect the energy away. Also, changes in aspect from missile or target maneuvering will alter returns as various parts of the aircraft deliver varying intensity ('glint'). This will produce 'jitter' in the processed signal and so guidance changes that have the potential for overwhelming the system. Varying atmospheric attenuation, target size, and stealth properties of the aircraft (specific outer moldline and inlet contouring plus radar-absorbent coatings) can further attenuate the signal. For multiple targets within the received return the seeker may fly to a mean value and so miss all of the aircraft. Since the SARH missile seeker is only a radar receiver and lacks many of the data processing features of an aircraft radar, it can be distracted by ground clutter returns, making consistent tracking of the target difficult in a 'look-down/shoot-down' scenario.

The power in the originating radar transmission, the radar mode selected (wavelength and PRF chosen), and sensitivity of the missile receiver plus signal process can help overcome many complications. However, factors affecting range means that doubling the detection range requires at least 16 times the emission power. Similarly, significant improvements in the performance of other aspects of the radar require large changes. At least advances in solid-state electronics and computer processing reduced the electronic 'noise' in the system and allowed for marked improvements with fewer fundamental changes. These have also reduced the tasks and proficiency requirements for the radar operator such that a dedicated crewmember became unnecessary.

While the illuminating aircraft can perform some maneuvering after SARH missile launch, it is important not to exceed the slew capability of the radar antenna such that the target is no longer 'painted' and RF energy is not reflected to guide the AAM during flight that could be in excess of 10sec – an excruciatingly long period in air combat. If the shooter must turn sharply away from the quarry owing to a threat (enemy fighter or missile) then lock can be broken and the shot spoiled. Intentionally breaking lock is fairly simple should the shooter suddenly realize that the target is a 'friendly.' With loss of lock the AAM usually continues in flight on its last course until momentum is expended and it enters a ballistic flightpath to ground impact, or a self-destruct function is tripped. It would only be in the rarest of instances that the missile strikes the target or approaches closely enough for warhead fuzing after loss of lock.

Active Radar Homing (ARH)

When radar had been reduced in size, volume and weight, enough to permit an entire unit installed in the missile, the active radar homing method became feasible. Improved reliability of radar components was also necessary for a practical weapon. Without the need for a fighter to continually paint the target for semi-active homing, the ARH is a launch-and-leave system (Figure 40). A fighter firing a SARH round at an opponent firing an ARH in response would be at a disadvantage in having to minimize maneuvering to maintain radar lock on the target while that aircraft can maneuver away while its fire-and-forget AAM continues homing. The ARH is also considered less susceptible to jamming than SARH systems given the stronger guiding signal strength and so the counter signal must be all the greater.

A principal limitation of an ARH seeker (Figure 41) is the size of the radar antenna and its gimbal limits in the nose of the AAM. The smaller the antenna the lower the emitted energy and so detection range. With desire to keep the AAM body diameter low, any antenna is much smaller than that of a fighter aircraft. The smaller the gimbal limits the lower the off-boresight capabilities. Improvements in these factors plus higher scan rates have been continuously sought to improve performance. The ARH also demands greater power generation within the missile than SARH. As with the aircraft radar in SARH applications, substantial improvements in performance requires large changes in system capabilities to include 16 times the power output to double the detection range – a tall order given the small packaging within a missile. Consequently, for a LOBL scenario, some SARH missiles initially had greater range than ARH.

In the 1990s radar-guided missiles began to appear with flat (planar) array antennas (Figure 41). The most common version in AAM applications is still gimbaled and has a series of slots across its surface. This slotted array antenna has waveguides behind directing the radar energy pulses (for pulse-Doppler) to the slots, dimensions of the slots matched to the transmission wavelength.[14] Altering the phase and amplitude of output at each slot forms the radar beam. The return is sampled through the same slots and waveguides. The passive phased array systems steer a single beam at a single frequency. A fixed planar antenna employs dozens of separate transmit/receive elements arrayed across the surface. These can transmit at slightly different frequencies and/or phase delay to shape and steer the beam without moving the antenna. This active electronically scanned array (AESA) system requires computer control and processing of returns. Application of this technology to AAM was just emerging at time of writing.

Electronic micro-miniaturization and more powerful rocket propellants allowed kinetic range beyond the LOBL range of the ARH seeker. It became possible to fire the round on a course determined by the aircraft radar and FCS with the AAM radar on standby until within detection range to detect the target and execute the endgame engagement – a LOAL mode. To ensure the AAM arrives at the correct spot in space to achieve LOAL for a potentially maneuvering

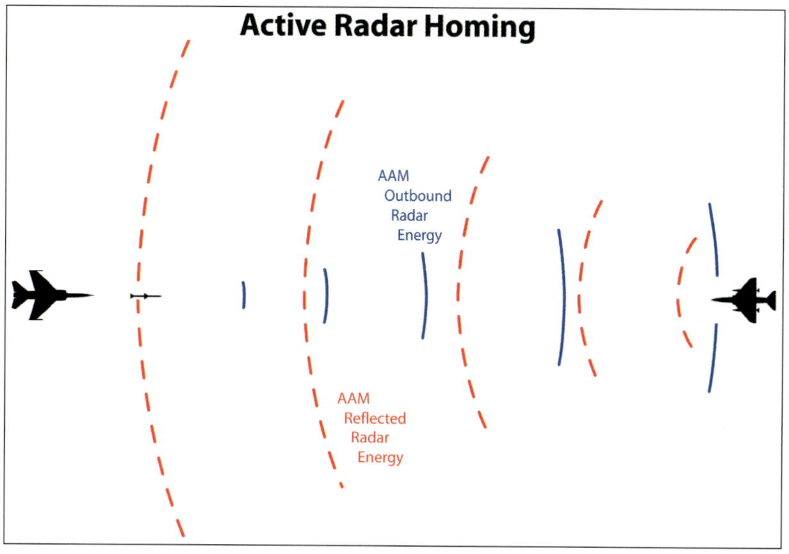

Figure 40: A depiction of the basic function of an active radar homing AAM shows the original radar emission from the missile and reflected back from the target. (Drawing by author)

Right: An example active electronically scanned array (AESA) radar antenna is seen as a fighter-size unit in a laboratory setting. It has hundreds of transmit-receive elements and need not be gimbaled to scan airspace ahead. (Northrop Grumman)

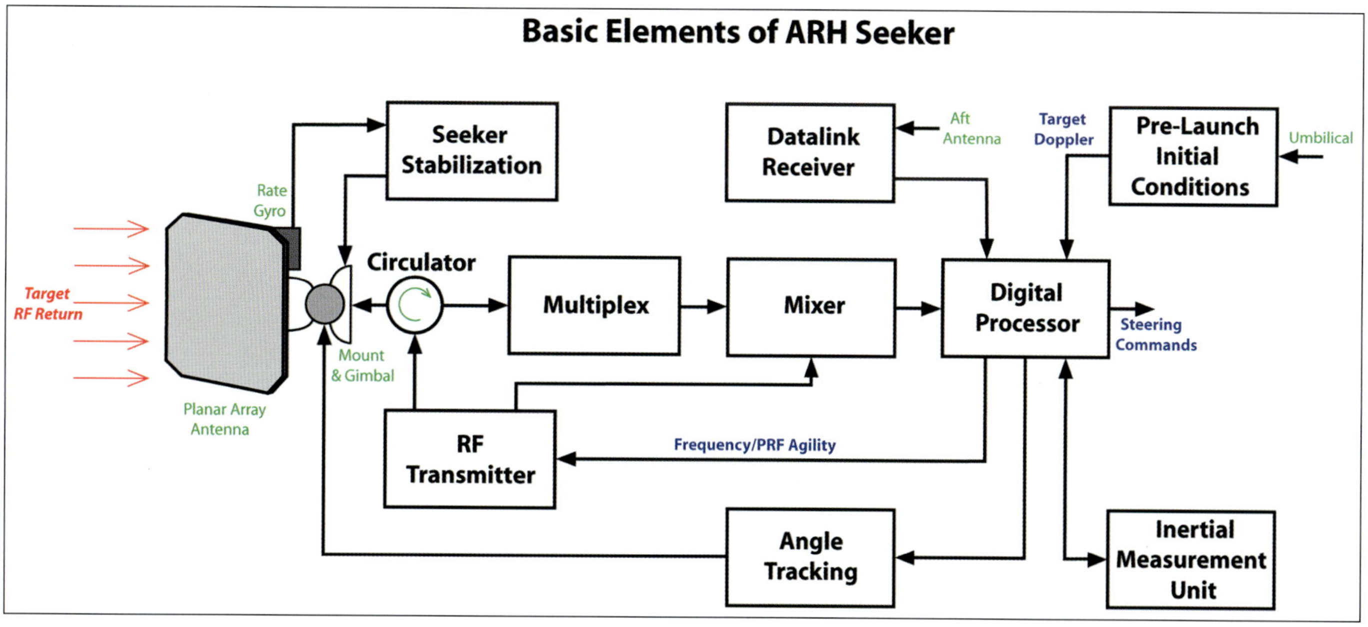

Figure 41: The basic elements of an advanced ARH missile guidance section is shown. Less-advanced systems might exclude a datalink receiver and substitute a more basic signal processor and a concave antenna. The Circulator turns the transmitted signal and the receiver on and off alternately. (Drawing by author)

target, the course can be set prior to launch and the antenna oriented to acquire the target when within range. If the targeting data is from an off-board source such as an AWACS, it is possible to engage targets even beyond the range of the launch aircraft's radar. Alternatively, mid-course guidance updates can be sent from the fighter's FCS via datalink transmission, received via an AAM aft antenna. This demands that the target be 'painted' by the launch aircraft for most of the missile's flight. Consequently, like SARH, the aircraft radar must continue tracking the target and its aircrew might detect this via RWR. Conceivably another platform can provide this illumination and datalink signal. Using satellite communications (SATCOM) to relay or 'bounce' the signal supports VLRAAM where the round may be beyond LOS from even an airborne emitter or beyond the curvature of the Earth. This also allows the AAM to be fired in a lofted trajectory maximizing the kinematics of the missile for the greatest range practical (Figure 42). Other means of 'trajectory shaping' and 'modified proportional navigation' have been employed to extend range without making a change to the motor or drag of the missile. The datalink undermines the launch-and-leave advantage of an ARH missile. It also represents another transmission that can be jammed or, with sufficient knowledge, 'spoofed' to redirect the AAM. If the missile arrives at an incorrect point for its radar to acquire the target, it may not be found and the shot is wasted. Table 7 is a list of potential variables contained in the datalink signal.

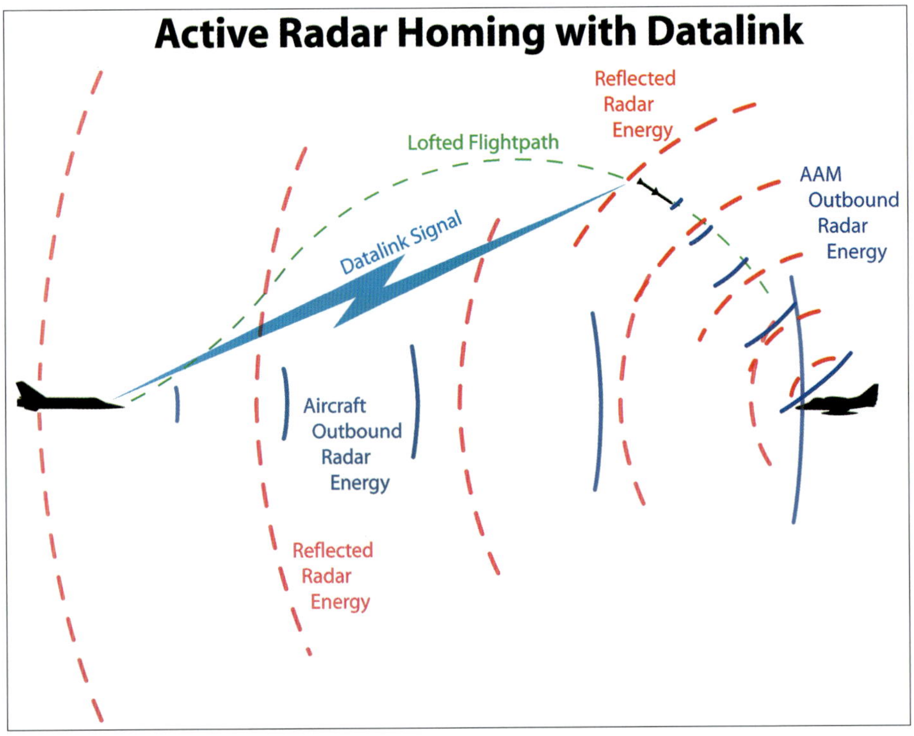

Figure 42: An active radar homing AAM aided by a datalink from the launch aircraft, or potentially some other source, allows exploitation of the full kinematics of the missile. This can permit the efficient lofted flightpath shown that extends range. The AAM's radar only becomes effective for endgame engagement when within its range (less than the aircraft's). (Drawing by author)

Table 7: Example Datalink Content[15]

Variable	Function
Target Position	Updates missile estimate of target position. Used in determining missile-to-target range, active acquisition initiation, and antenna positioning
Target Velocity/Doppler	Update of Doppler filter positioning
Active Acquisition	Used if it is the function of aircraft avionics
Launch Aircraft-Missile Velocity/Doppler	Doppler update for system with sample data semi-active mid-course guidance
Angle Update	Positioning of missile antenna if missile requires computation from aircraft avionics
Missile Identification	Defines which missile the particular datalink message is for if several in flight simultaneously

The greatest challenge for the radar-guided missile, SARH or ARH, has been the ability to detect and track targets below and so against surface background. The difficulty with radar return clutter in tracking low-level targets in look-down/shoot-down remained a technological challenge for decades. Only the advent of digital processing via programmable microprocessors greatly reduced this disadvantage.

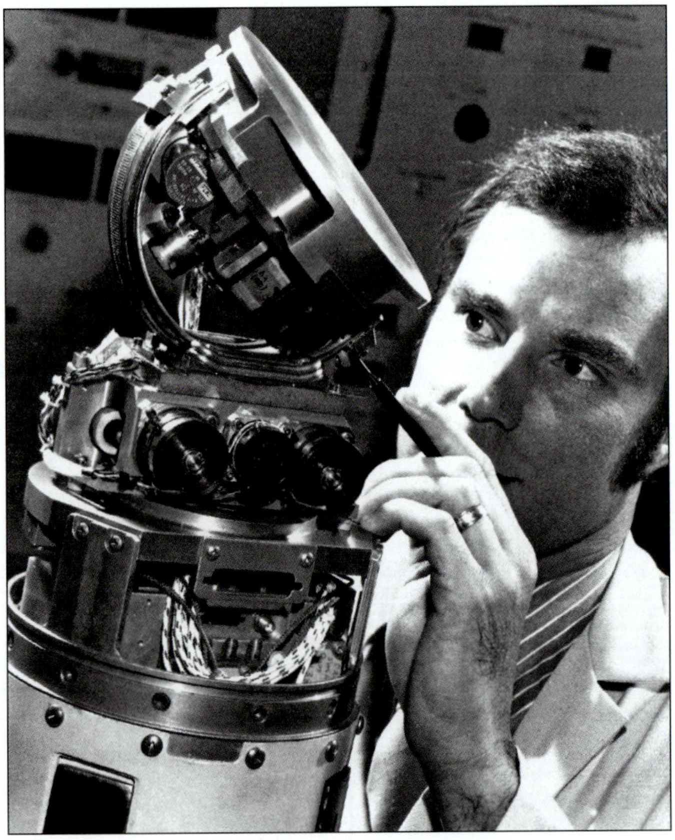

The 'guts' of the ARH missile is suggested by this image of a developmental unit from Hughes, offered during the AMRAAM competition (Volume 4, Chapter 1). The flat antenna suggests the new phased (planar) array radar technology. (Author's collection)

The ARH ideal – an AMRAAM missile executes a forward quarter aspect engagement of a QF-100 target drone. A feature allows it to delay detonation until passing a bit along the fuselage length. The explosion shows the annular explosion typical of AAM warheads. The drone does not appear to be maneuvering and likely not emitting electronic countermeasures signals. (National Naval Aviation Museum)

Most radar-guided missiles with a pointed, solid nose have a metal tip as shown in these AIM-120s installed on an F-16. These are to prevent rapid erosion of the nose cone material with the impact of raindrops at 2 Mach or greater. (USA DoD)

Infrared Homing (IRH)

This 'heat-seeking' AAM has been the principal passive dogfighting missile with engagements WVR of the attacking pilot or initially approximately 3mi (5km). Especially jet-powered aircraft have a very hot exhaust (hot gases, heated exhaust nozzle, and super-hot final turbine stage forward of the nozzle) with associated emissions in the IR spectrum. Once launched the IR-guided AAM scans to track the target entirely autonomously (passive homing, Figure 43) and the attacker can turn away to evade closing adversaries or inbound missiles. This is another fire-and-forget system and detecting them inbound is most challenging. Heat-seekers especially support night intercepts – though not positive target identification.

The intensity of the exhaust IR emissions drops off as bearing increases away from the aft centerline of the aircraft (angle off the tail) or aft of an engine exhaust for a multi-engine aircraft. The IR energy is focused in the AAM seeker head onto infrared detector material that alters any current passing through it. That change in current is amplified and used in guidance electronics. Initial IR detectors were not sensitive enough to permit engagements much beyond directly aft, and so a tail-chase intercept remained the most practical for several decades – although employing proportional navigation that provided lead. Also, with a high angle-off-the-tail,

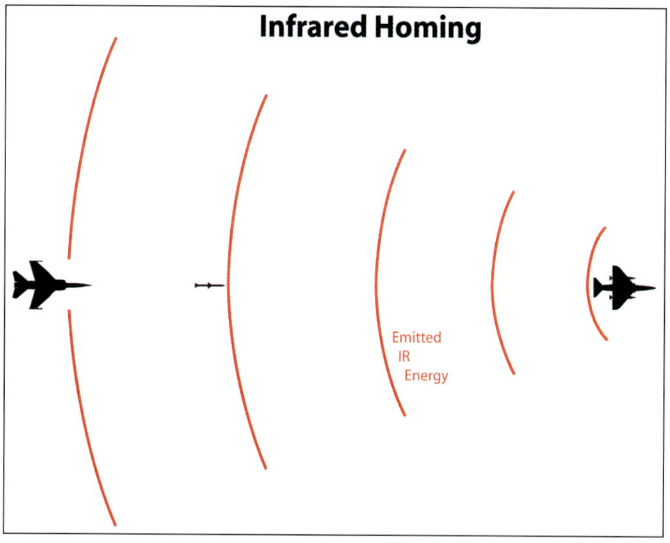

Figure 43: A depiction of the basic function of an infrared homing AAM shows the IR emissions of the target aircraft radiating in a manner that can be detected by the missile seeker head to guide the round autonomously. The IR signature is actually a full sphere around the target, but is most intense in the rear quarter where the engine exhaust is clearly visible. The IR energy can also be detected by launch aircraft sensors, if so equipped, to cue the seeker. (Drawing by author)

the AAM has to perform greater energy-bleeding maneuvers during the intercept, cutting range. If the target aircraft turns into the missile in an effort to break lock, moving rapidly across the missile's flightpath and possibly leaving the seeker FOV, the AAM has to maneuver even more sharply – potentially beyond its remaining capabilities. Lead pursuit and lead-collision trajectories (Figure 24) also require greater and faster guidance processing.

Airborne jet aircraft have multiple IR sources at various temperatures and emission intensities. However, the jet exhaust is the most intense. Figure 44 illustrates how the typical 500°C temperature of the jet engine exhaust metallic elements radiate solidly in the IR bandwidth, with the peak intensity at around 3.75 microns. An engine in afterburner would have a peak around 2.0 microns (Figure 44). The jet plume, extending 20–50ft (6–15m) behind the aircraft, itself is composed to a great percentage by carbon dioxide and water vapor. These radiate at about 3 microns for the gas and 4.2 microns for the vapor and, although spreading to a wider source than the exhaust (roughly 2–4 microns for the metal with afterburning and 4–5 for the plume), it cools rapidly as the aircraft moves way (Figure 45). The exhaust is invisible from the aircraft forward quarters and little of the plume is visible. Internal components, such as the aft face of the turbine and the afterburner flame-holder, can be 1,000°C and, if visible to the AAM seeker, is an intense IR signature, albeit of narrow aspect looking up the tailpipe. Turbofan engines, common in fourth and fifth generation fighters, and turboshaft engines in helicopters, have less intense emissions and exhaust plumes extending approximately a tenth as far as turbojets. Skin friction for an aircraft at supersonic airspeed can raise surface temperatures by 150°F (65°C) and radiate at around 5–13 microns bandwidth (leading edges at 5.5 to 8.6), and so of much less intensity as suggested by Figure 44. The totality of the IR emissions from the aircraft is suggested by Figure 46.

The sensors have been designed to the extent permitted by available technology to react only to spot IR sources and those in a narrow bandwidth associated with engine exhaust and heated aircraft structure. However, background and environmental IR sources can still confuse the guidance. The best launch condition for the IR detector are the target against the sky with the sun to the rear. Worse is against the sun, as a major IR 'emitter', serving as a potential source distracting the seeker. The sun's IR energy is also reflected by atmospheric and environmental elements, particularly reflections off clouds and, in look-down, water, snow and sand. Hot point ground sources and sun glints off windows can also confuse the sensor. If any of these present a greater IR source than the target aircraft, then the seeker can be diverted and lose lock on the aircraft during its scan. Sun reflections off the aircraft (glint) are a function

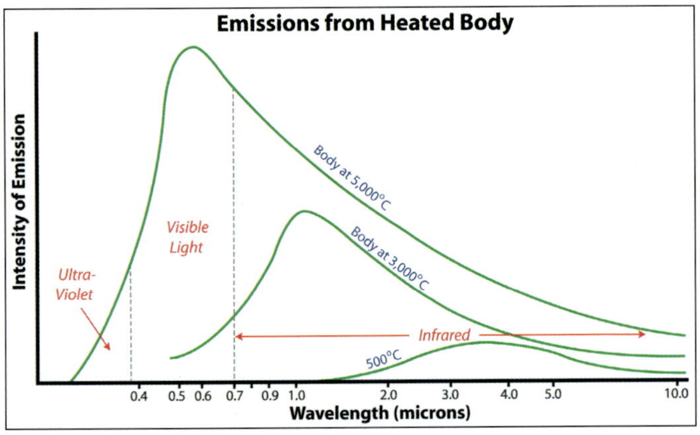

Figure 44: A body at varying temperatures will radiate electromagnetic energy at different wavelengths and intensities. For the exhaust elements of a jet aircraft engine this is approximately 500°C where the energy falls in the infrared band, making this the best choice for an optically guided AAM seeker. The peak is at approximately 3.75 microns. (Drawing by author)

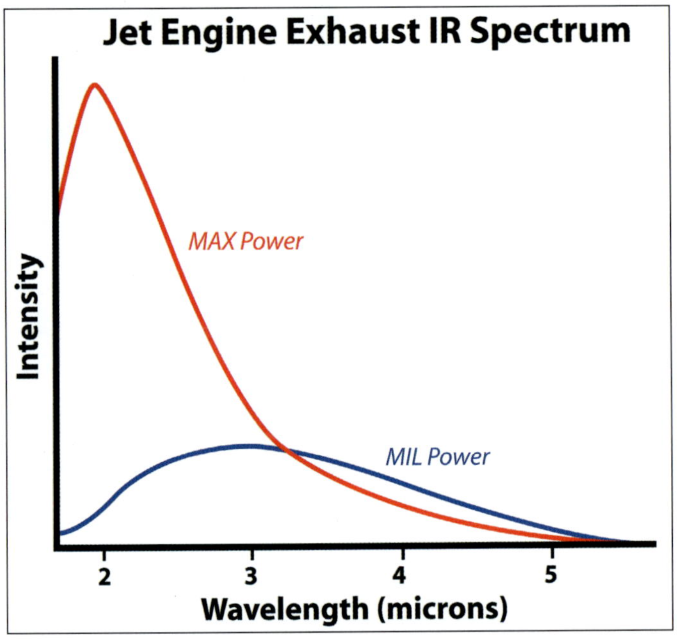

Figure 45: Where the jet engine exhaust falls in the micron scale and its intensity depends on whether it is operating at MIL power (Military, full throttle but short of the afterburner, AB, range) or MAX Power (full AB). An IRH seeker designed to encompass this entire bandwidth would find MAX Power an optimal IR signature source for guidance. Unless seeking to outrun the AAM (if detected), the targeted pilot would be well advised to pull the throttle(s) out of AB. (Drawing by author)

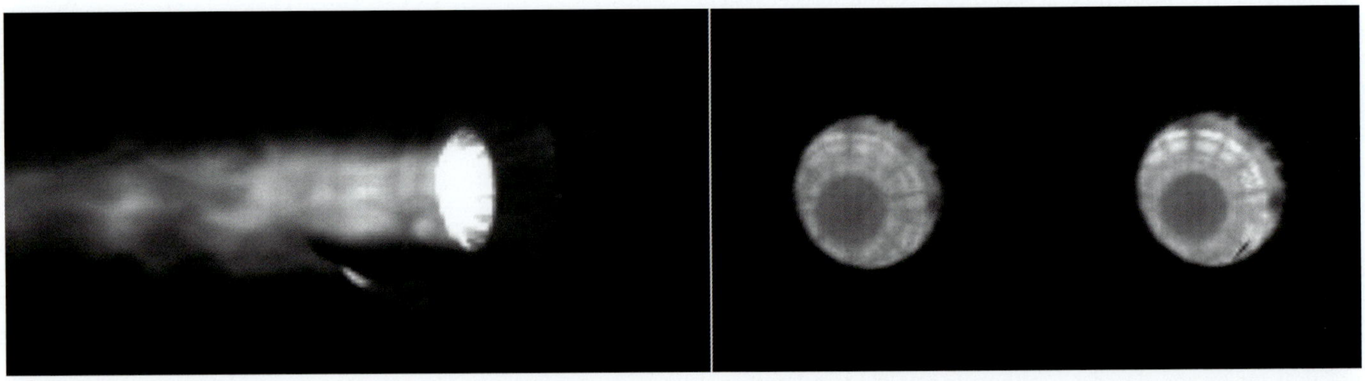

Infrared images of F-14A operating TF30 engines illustrate the exhaust plumes at left and the internal components at right that create sources for IRH missiles to track. (Public Domain)

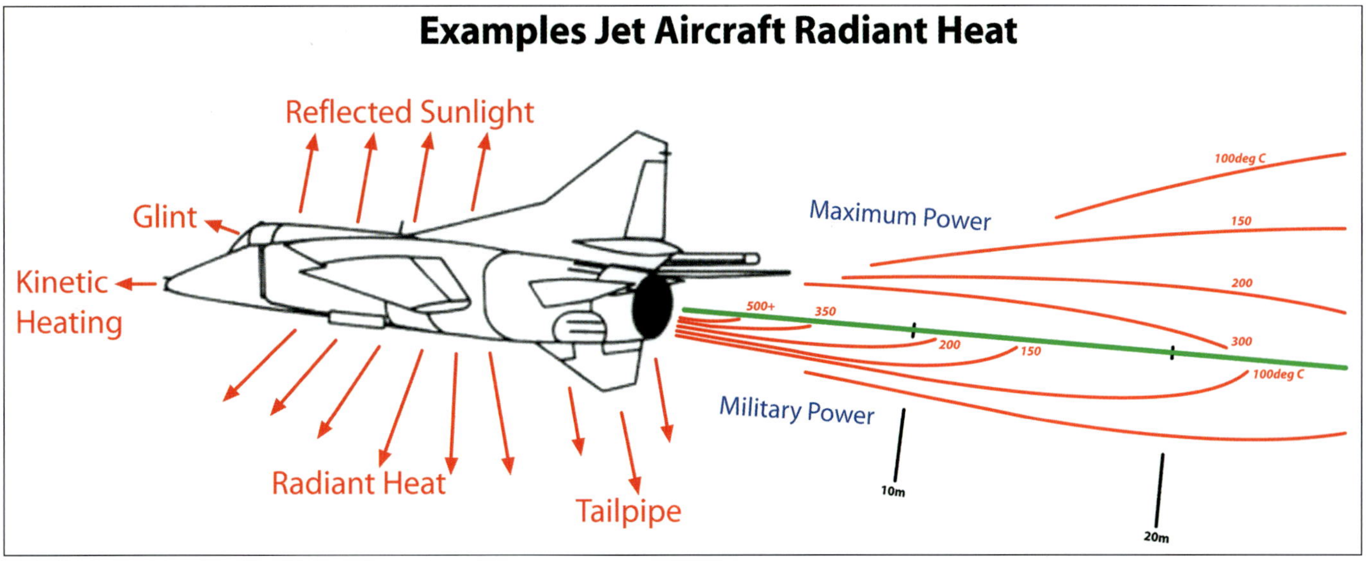

Figure 46: Shown is an example of the radiant heat from an aircraft (MiG-23 in this case). The heat within the exhaust plume is indicated as it extends aft and vertically, with afterburner on top and without reheat below the centerline. Note that the plume is wider in AB. The intensity of the reflected sunlight is a function of the aircraft finish and glint off the canopy a function of aircraft attitude with respect to the sun and incidental reflection into the missile's seeker. The radiant heat refers to that from operating internal components that would include the engine bay. The tailpipe is heated by the engine via conductance and the exhaust via radiant heating. Kinetic heating is from passage through the air. (Drawing by author.)

of the surface finish (dark and flat paint reducing this) and flat panel canopies (common in attack helicopters) give a glint in only one direction versus omni-directional for curved surfaces. Additionally, the IR energy is scattered by atmospheric factors such as water vapor, clouds, rain, smoke and haze to degrade S/N and greatly reduce the effective range of the missile, in some cases to virtually zero. Even in relatively clear skies the energy is attenuated over a distance to limit the effectiveness of IR guidance compared with radar. Just cold water vapor of varying droplet size and carbon dioxide in the air can absorb the transmitted energy. This is suggested by Figure 47 showing varying transmission 'windows' at varying wavelengths at sea level, but shifting with altitude and local atmospheric composition. Heated airframe emissions, for example, are less severely impacted at high altitude and the faster the target is flying. Hence, the selection of detector effective wavelength can be ideal in some circumstances but less so in others. Generally, rain and clouds have the greatest effect, less so smoke and haze. The IR emissions from heated airframe components in the 5–8-micron bandwidth is easily absorbed by the atmosphere while those at 8–13 microns are subject to less absorption but are also of less intensity. Only after decades of development of more sensitive seekers and more effective filters (see later) did these factors change. It then became practical to prosecute far more distant WVR engagement than dogfighting, up to approximately 5mi (8km). Even BVR attacks became possible, out to the kinetic limits of the missile.

The IR seeker is at the front of the AAM behind an optical window also making up the aerodynamic nose of the missile. These commonly use a Cassegrain type telescope (or Schmidt-Cassegrain) to focus the scene onto a small IR detector. This telescope has a spherical concave primary mirror that reflects the incident light of

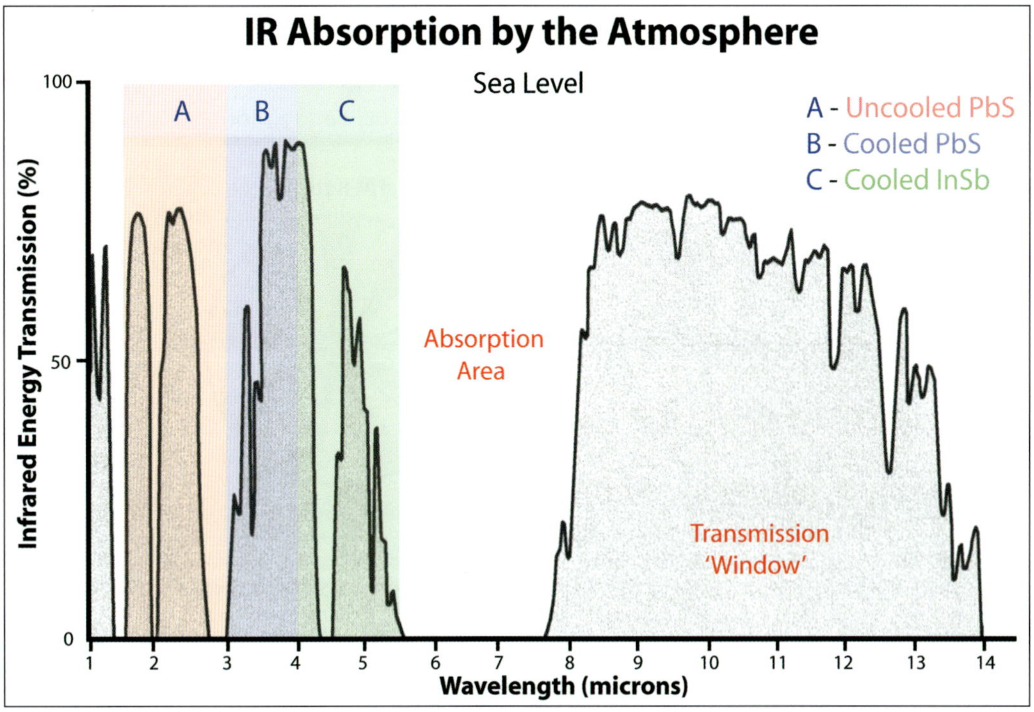

Figure 47: Infrared energy can be absorbed by water vapor and carbon dioxide in the atmosphere, degrading the performance of an IR seeker. Shown are various transmission windows separated by absorption areas at sea level. Suggested in this simplified representation is an ideal window of 8 to 12 microns, but most detectors in IRH seekers operate much lower in the band where most aircraft emissions occur. The effective bandwidths of three early detectors are superimposed. (Drawing by author)

the scene outside the transparency onto a convex secondary mirror that then focuses the light at a point beyond the primary, the light passing through a small hole in its center. The telescope greatly narrows the field-of-view, or the span of degrees that is 'seen', like the vision of the human eyeball without turning the eye. This helps to avoid detecting environmental IR sources. The instantaneous field-of-view (IFOV) is this narrow 'vision' of the seeker sensor during any instant in the scan of the field-of-regard and falling on the detector. The IFOV and can be a few degrees or as small as 0.1deg. To enable a good FOR (Figure 3), the telescope is gimbaled for off-boresight sweeps of the FOV, such as the limits of human sight by turning the eyes. The optical FOR can be reduced in the near field by obstructions like adjacent stores and the launch aircraft airframe.

The optical assembly is generally free spinning to be essentially a gyro, inertially stabilized and so decoupled from the missile body motion. Early designs had permanent magnets on the spinning sensor gyro that passed through an electromagnetic field generated at the surrounding missile body to produce an alternating current (AC, a sine wave) serving as an angle reference. Other means of doing this emerged later. Early generation missiles also had the seeker 'caged' (fixed) on boresight until after launch to avoid damage in the tremendous acceleration of firing. This meant that the launch aircraft had to be maneuvered to place the target roughly in the center of the aiming reticle projected on the gunsight such that the seeker detected the target within its limited FOV. Later missiles could be launched with an uncaged seeker and so had a larger initial target acquisition angle. The missile indicates to the

An AIM-9 is posed as if for launch (the round is inert) from the side weapons bay of an F-22A. The design of the bay and extendable launcher places the missile at a substantial angle to the aircraft's flightpath and so off-boresight even before launch. One side hemisphere is blocked by the bulk of the fuselage, but only for very close shots (and possibly inside minimum range). With a lock-on-after-launch capability, this need not be a deficiency. (Author's collection)

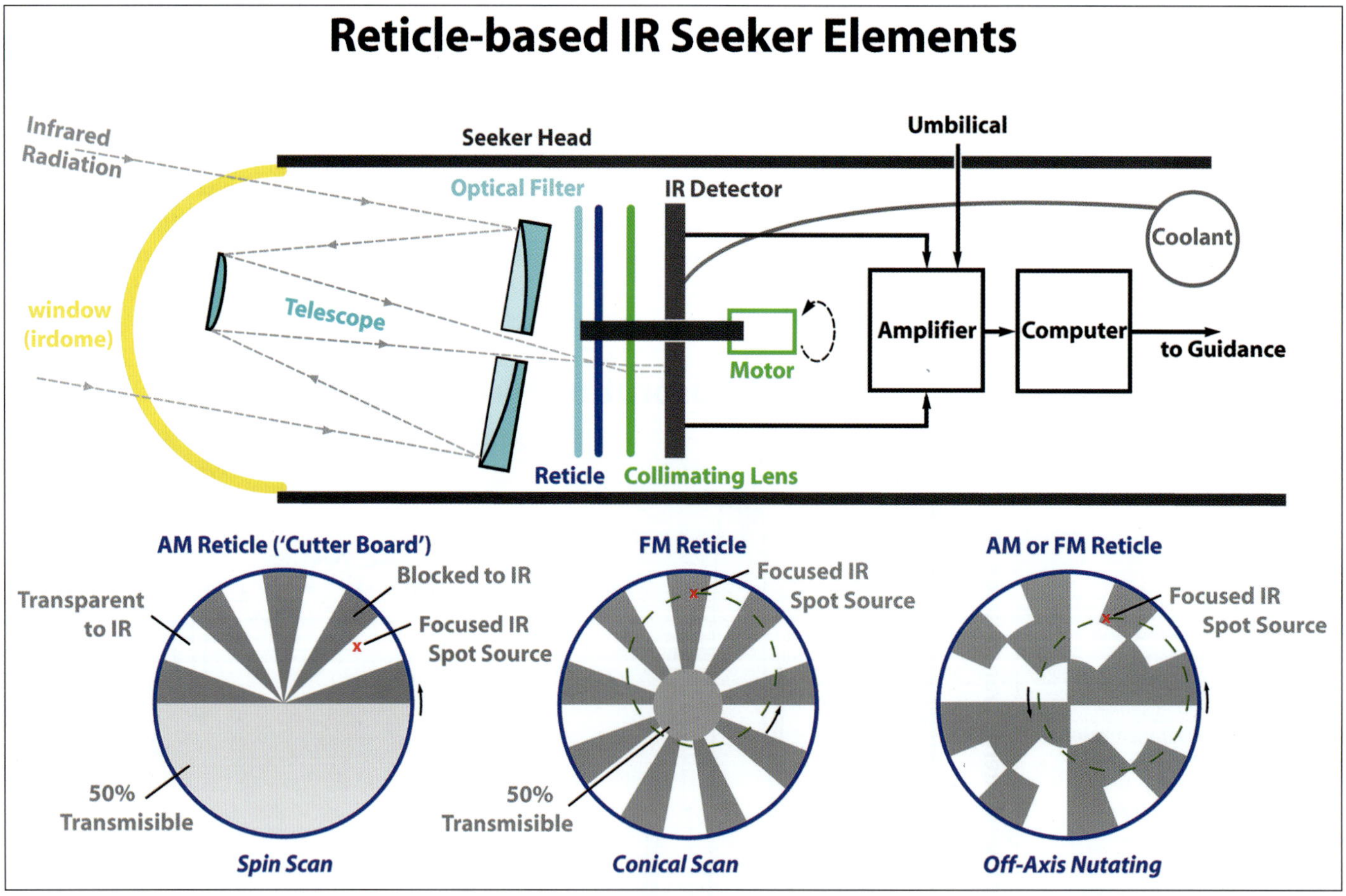

Figure 48: The drawing depicts the reticle-based (non-imaging) IR-guided missile seeker head with many moving parts. The initial amplitude-modulated 'cutter board' reticle of the spin scan technique is shown with the later frequency-modulated conical scan reticle. An even later design accommodated the off axis nutating scan and, depending on the reticle pattern, could be AM or FM. The IR energy impinged on the detector that delivered angular information to the guidance system. Note that the second reticle design represents a non-rotating version. The third from left is better able to naturally filter large horizontally oriented IR sources like clouds and lakes. Some such reticles had a denser checkerboard. The detector coolant system was not evident in the earliest IRH missiles. (Drawing by author)

pilot once it has acquired a point IR source, generally via a tone in the helmet earphones and possibly a cockpit light. These could have varied intensity indicating the quality of the track, the pilot firing at lock-on.

The first few generations of IRH missiles employs a reticle that intermittently interrupts the IR point source energy reaching the detector and naturally filters broad natural sources. As the gimbal sweeps the telescope's FOV across the FOR, the incident light passes through the reticle and is focused on the detector. As shown in Figure 48, the circular reticle is a baffle ('chopper' or 'cutter board') with half the alternating sectors open to pass the light to the detector and the others (typically aluminum deposits on glass) blocking the light. The reticle spins with the gyro, generally at 75–200 cycles per second (Hertz, Hz). A typical reticle is around 0.5in (1.3cm) in diameter or less. The layout of the sectors and the rate of rotation serve as a natural filter, modulating the light passing through and narrowing the IR bandwidth to help in tracking desired IR targets rather than environmental sources. A large IR source such as the sun, or sun reflections off clouds or snow, would fill the detector (pass through more than one sector simultaneously) and so not be interpreted as is a point source but generating a steady output that can be electronically filtered. A point source like a distant jet engine exhaust would be much smaller and so periodically obscured. The reticle intermittently interrupting the point light source reaching the IR detector means that the detector's output is a modulated AC current which is passed through an amplifier to the guidance system which is designed to react to the signal. The frequency of the output current (Figure 29) is a function of the spin rate and the number of transparent sectors.

When the IR spot is off the center of the reticle the energy passing across several sectors is interrupted in different ways before reaching the detector. This creates a difference in IR intensity measured in the areas of the detector where the light falls. The output has a higher rate but lower intensity closer to the center of rotation than further out in radius. A comparison between the output AC signal and the sensor angle reference signal allows vertical and lateral guidance information to be derived. The information is used to direct the FOV scan within the FOR to center the target on the reticle (equal intensity across the detector) such that the two signals are coincident. The earliest systems passed the current from the detector, through the amplifier, to electrical solenoid coils surrounding the gyro. This caused the gyro to precess and track the target. This is effectively a natural proportional navigation system, maintaining a constant LOS in homing on the IR source (Figure 25). Lacking the coils, other processing and precision motive devices are necessary. When the spot source is centered on the reticle (boresight), gimbal scanning stops at that angle and the telescope spins in a small circle ('nutates') so that deviations can be detected with finer resolution. This arrangement is called spin scan. The consequence of this design is that new guidance commands are only generated when such displacement occurs, and the missiles tends to follow a snaky flightpath as it repeatedly deviates and corrects. As the missile approaches the target, especially with the early bang-bang control systems, the control response can be coarse (over-control) and produce a miss in addition to draining energy for a shorter range. However, with navigation seeking to lead the target, the possibility of it being on boresight is reduced. Additionally, as the point source grows in size until it exceeds a sector field where no modulation for guidance. Similarly, an IR countermeasures flare (Chapter 4) dispensed by the target aircraft will move out of the null center and can pull the tracker off the target.

As seen in Figure 48, several possible reticle layouts have been employed with other distributions of sectors, and more not shown. The earliest designs, reflecting work by Germany and Eastman Kodak in the USA for guided bomb during the Second World War, used a reticle like a spinning fan with four 'blades.' The layout shown at left was used in the first fielded IR-homing AAM, the Sidewinder A (Volume 2, Chapter 2). The 50-percent transmissible section outputs a direct current (DC) that, when subtracted from the AC signal, helps to reduce the influence of broad IR sources. This also plays a role in determining direction to the point source target by outputting an amplitude modulated (AM) signal from the detector for comparison with the reference waveform amplitude for the magnitude of the tracking error. The full fan design at center in the figure is more modern and effectively a frequency modulating device (FM, comparing waveform frequencies, Figure 29). This also assists in infrared countermeasures rejection (IRCM, Chapter 4). The FM system has generally been regarded as superior to AM as it has better signal quality and experiences less noise interference.

Figure 48 also shows common scanning methodologies for modulating the IR spot energy reaching the detector. The first was produced by the rotation of the reticle for so-called spin scan. In the second the telescope's secondary mirror is tilted slightly off axis and so, with spinning, the image focused on the fixed reticle spins (nutates) in an offset circle. The result of this conical scan ('con-scan') is that the point IR image is never fixed at boresight – at best describing a circle around the null point – and so overcomes the null center limitation of the AM reticle and helps in avoiding decoys since it is never 'staring' at the target. A flare would leave the FOV quickly and so is less likely to pull the seeker off the target. The optics are generally designed to spin the target close to the edge of the reticle for the greatest FM variation and so tighter tracking. Where it leaves the reticle the potential for jamming from the target is reduced. Flight tends to be smoother, with less energy-draining undulations. Other designs employ both AM and FM features, such as inclusion of a 50-percent transmissible circle centered on boresight shown in the center reticle of the figure. More exotic were detectors in various patterns (far right example) and scanning techniques offered different advantages, especially when employing digital processing (prior systems being analog). One benefit was being able to track rate of change of the spot energy on the detector and so anticipate the next motion as a means of rejecting divergent signals presented by flares. While more were studied, the rise of non-optical imaging systems (see later) meant that these were ultimately set aside.

The initial detectors were photoconductive, the IR energy altering the material's electrical resistance and so any current passing through it. Later these became photovoltaic, the IR energy directly generating current from the semiconductor material proportional to the incident IR energy. The first generation of AAM employed photoconductive lead-sulfide (PbS) IR detectors, the material bonded to a base like a ceramic wafer. This is sensitive to the near-IR (short wavelength) bandwidth of about 1–3 microns where IR intensity from a jet engine's hot turbine is good, less so the exhaust plume, but also sensitive to the sun and reflection, and where there is an absorption window (Figures 44 through 45 and 47). This implies a tail-chase engagement against a clear sky and without the sun in the FOR – significant limitations. Second and third-generation missiles introduced different detector materials (Table 8) and photovoltaic systems. These expanded the bandwidth to higher values such that heated elements of the targeted airframe were detectable – reaching for an all-aspect capability. The IR energy reaching the detector can be altered with a glass interference filter (Figure 48), or several stacked, to narrow the bandwidth somewhat and reduce confusion with reflections. The transparent nose of the missile can also be treated to provide such filtering. There are absorption filters that cut out large bandwidths of energy and interference filters that pass narrow bandwidths with good transmittance. Stacked layers of filters have varying thickness owing to reflections off the surface of each layer and refraction within the layer that will affect frequency transmittance. With the lower intensity emissions and absorption potential of higher bandwidths the detector must be more sensitive. Apart from selection of detector material, it can also be cooled to increase sensitivity (Figure 49). Cooling the detector can raise the effective bandwidth cutoff by approximately 1 micron. A photovoltaic indium-antimonide (InSb) detector chilled to below -150°F (-100°C), for example, is good for 3.2–5 microns for a reasonable all-aspect capability. That is 35 percent of the IR spectrum versus the uncooled PbS seven percent.

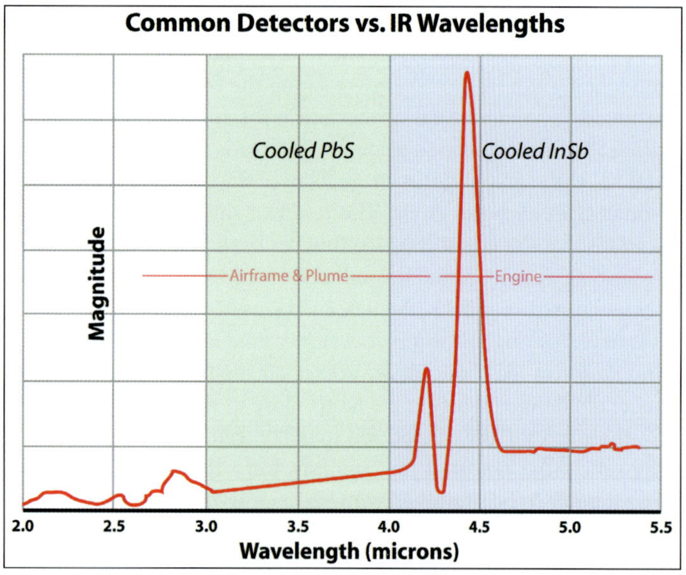

Figure 49: The benefits of more capable IR detectors via cooling is illustrated here, spanning bandwidths encompassing more of an aircraft's infrared emissions. (Drawing by author)

Table 8: IR Detector Material Characteristics

Material	Identifier	Bandwidth (microns)
lead-sulfide	PbS	1–3.2
indium-antimonide	InSb	3.0–5.5 cooled
lead-telluride	PbTe	4–5
lead-selenide	PbSe	1.5–6.0
mercury-doped-germanium	GeHg	2–14
mercury-cadmium-telluride	HgCdTe	0.8–14*
indium-sulfide	InS	?

* Range dependent on chemical 'tailoring' and cryogenic cooling

Detector cooling has been done with carbon dioxide (CO_2), liquid nitrogen, and argon to bring the material's temperature down as much as 390°F (200°C). This has used a Joule-Thompson gas expansion ('throttled') process by which the gas is passed through an orifice or valve under pressure. Upon expanding past the orifice, a drop in temperature occurs. More efficient is the Leidenfrost process in which a drop of liquid coolant is exposed on the detector, 'boiling off' a portion to gas that sustains the droplet above for a bit longer to conduct away heat from the detector. Another design is the Stirling closed-circuit system in which a coolant gas is compressed and circulated to collect heat which is then dumped through expansion to a larger volume and a radiator before repeating. Thermoelectric cooling, without a gas or liquid, typically employs the Peltier effect. Cooling occurs when a current is applied to two dissimilar materials bonded together, transferring heat from one side to the other. This does not cool as much as with a refrigerant and functions only as long as the current is applied (up to battery depletion during free flight).

One early concern with the heat-seeker was that the IR seeker would only go for the engine exhaust plume and possibly detonate short of the aircraft without lethal effect. This could be especially true when the adversary pilot selected afterburner. The solution was to tailor in lead bias by which the missile aim point is adjusted forward of the initial guidance command – an added lead to the proportional guidance command – or 'rate bias' that aims at the most forward heat source within its bandwidth. From that point the fuze determines when the warhead is detonated (see earlier).

A major change to the IRH missile seeker was sought by employing multiple small detector elements to produce a crude image of the scene for digital processing and more effective guidance and IRCM rejection. Maturation of the technology, along with the digital processing capacity and speed, and then making these affordable, delayed the implementation of this seeker type. As this imaging infrared (IIR) technology matured, a first step were hybrids employing reticle-like features ('pseudo-imaging') with detectors directly on the disk that may or may not be spun but onto which the image is scanned optically. Two examples of these are shown as the first two arrays from the left in Figure 50. These can also employ two different detector materials sensitive to different and narrow IR bandwidths ('two-color') for enhanced all-aspect engagement. It may include a portion of the UV spectrum. This can assist in discriminating flares emitting in one band. These have been, as an example, designed to detect both the lower wavelength emissions of an engine exhaust as well as the higher emissions of airframe heating to provide all-aspect detection and intercept while also avoiding IRCM and environmental sources. The crossed reticle (crossed-array tracker, CAT) with con-scan is superior in flare rejection when matched with digital processing in that it anticipates movement of the target and so deviating sources like slower-moving flares falling away are rejected. In some cases, the primary mirror in such seekers is fixed and only the secondary nutates. Adding another mirror or a prism, counter-rotating, can create a rosette pattern (Figure 50) – the lobes scanned in a spinning pattern. This scans more of the FOV more rapidly and assists in discriminating multiple IR sources. As this would include more environmental and IRCM sources, digital processing becomes critical for discrimination and a must in extracting position data for guidance computations. Such processing can also predict the path of the target being tracked and only accepts output from the detectors in the time window at which the target should appear or rejects signal outside the window. This serves as another counter-countermeasures (CCM) and environmental source discrimination means.

By the twenty-first century the IIR sensor had become a focal plane array (FPA) of many light-sensitive elements, like a square of 128x128 sensors (16,384 total) arrayed on a circuit board 1in (2.5cm) square behind optics and with cooling still highly beneficial. Typically, the detectors are InSb material cooled to -600°F (-350°C). This can be a fixed or 'staring' FPA, not required to spin to scan the FOR, and so exposing a greater detector area for a longer period to the target's IR emissions. The optics are typically still gimbaled to move with the telescope for its limited FOV to cover a suitable FOR. Filters might still be used to eliminate off-band radiation to simplify

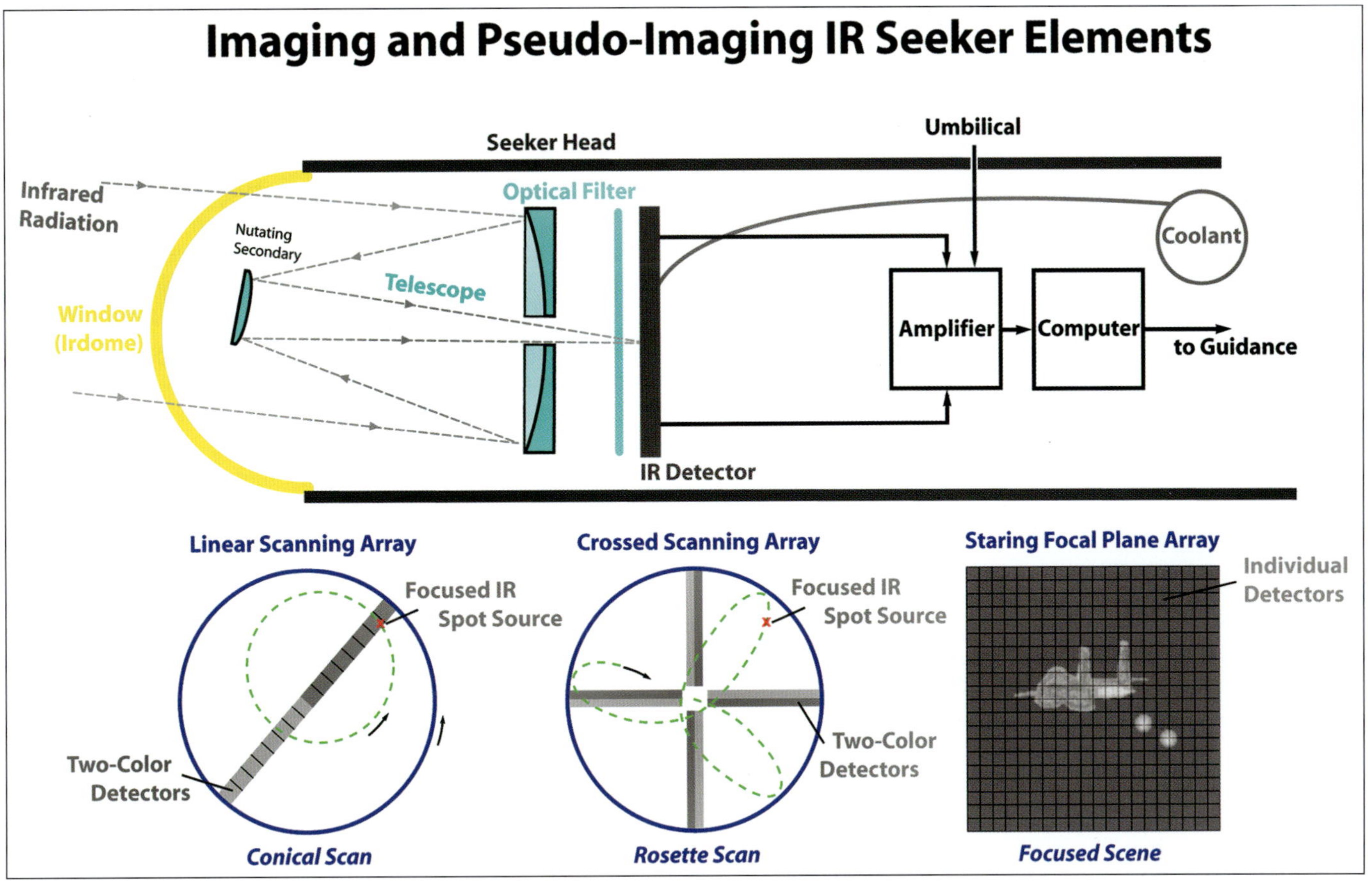

Figure 50: The imaging-IR seeker does not use a reticle but focuses the image directly on the IR detector that may or may not spin. Early examples employed a linear array of detectors or a crossed pattern, these offering the opportunity to use detectors sensitive to different parts of the IR spectrum. Various scanning patterns are used, these typically accomplished with a nutating secondary mirror (tilting and rotating), the primary fixed. In such cases the detector did not need to rotate (motor not shown in drawing). When the focal plane array became practical and matched with digital analysis of the IR scene within the FOV, a fundamental change occurred with the IRH missile. The optical filter shown is not universally used. (Drawing by author)

analysis. These are imaging systems in that the entire scene in the FOV are detected (Figure 50) and IR sources isolated via digital processing of the pixels defined by the individual detectors in the array. At 6mi (10km) range an aircraft may be just one pixel, but this grows to several then many pixels as the range closes. A flare, by contrast, will likely remain a single pixel – albeit at high IR intensity – aiding rejection. Even if the flare image encompassed several pixels, the image would not match that expected for an aircraft. A multi-spectral array uses a pattern of detectors sensitive to different bandwidths (colors). Software determines the choice of IR signal on which to guide. The array of sensors is rapidly sampled digitally in a line-scan process (sequential horizontal lines) like a television and video raster scan that, theoretically, makes it less vulnerable to flares and blinking decoying (Chapter 4). Data processing can also improve the rejection by assessing source intensity rise and decay time plus comparison with predicted target trajectory. As the AAM approaches the target and it begins to fill the array, comparative algorithms can be used to identify the aspect angle and adjust guidance for endgame intercept at the most vulnerable area of the aircraft or missile.

Thus example of a focal plane array has thousands of individual light-sensitive element. It likely measures about an inch on a side. (Public Domain)

This is an IIR focal plane array image of a QF-4 target drone as 'seen' by an inbound AIM-9X missile in the final split second before impact. It shows how the seeker perceives the scene and how, with scene comparison algorithms, the most vulnerable area of the identified aircraft can be specifically targeted. (Jim Rotramel collection)

The linear array of, for example, two rows of 64 elements can have some superiority over a FPA. It may be less prone to errors from vibration and irdome heating. The elements' sensitivity, or dynamic range, can be greater because of a larger area to store the charge. This can allow it to detect weaker targets. The array can also be cooled more quickly and require less refrigerant gas. It can also be expected to cost less to fabricate.

The IRH seeker lays behind a transparent dome wide enough to accommodate the gimbal and FOV limits of the optics. Glass does not pass much IR energy so more exotic materials are adopted, machined or cast then polished to optical quality, to create the 'irdome.' The material or coatings can be selected to filter certain IR bandwidths. Selected for these applications have been arsenic, arsenic-trisulfide, germanium, magnesium fluoride, quartz, sapphire, silicon, and trisulfide. Most of these are not clear to visible light and so appear to the eye as milky, yet pass IR energy. The dome is given a scratch-resistant coating, but still needs to be protected to avoid deep scratches that could scatter the light. This is commonly done with a cover that is removed before flight. However, the dome can degrade over many flights owing to the impingement of particulates in the air and bug impacts. Moisture condensation and ice accumulation would inhibit effective operation. Heating of the irdome during high-speed flight, sometimes unevenly, creates scene distortions to lower S/N and so reduce sensitivity to low intensity radiation. Heat generated within the seeker head can also degrade performance. Shockwaves forming and moving on the irdome, with varying air density, can further distort the image. In later decades, with onboard digital processing, computational corrections for this distortion became possible.

The optical window of IRH missiles (AIM-9E inset, opaque to visible light) must be protected from even the smallest scratches and pitting to ensure against light scattering and degraded seeker performance. A cover during ground handling and prior to flight is common. This is shown with an AIM-9E on a USAF F-4E aircraft. Note the wrap-around Velcro cover, with REMOVE BEFORE FLIGHT streamer, over the proximity fuze sensor windows farther back on the body. (Jim Rotramel collection)

With more IR bandwidth, narrow two-color detectors that can include UV, cooling for high sensitivity, the ability to track friction-heated airframe elements and sun glint, improved optics with uncaging prior to launch accommodating cueing, HOBS and high scan rates, all matched with digital processing, the all-aspect IRH with excellent discrimination became practical. By fourth-generation missiles the 'no-escape zone' (NEZ) engagement envelope – where it can out-maneuver any evasion by the target aircraft for a suitable intercept – was greatly expanded (Figure 51). Higher scan rates reduce the quantity of IR energy reaching the detector and, in late-generation systems, lower rates were acceptable given other design aspects. The HOBS also means that the optics will not optimally focus the energy on the detector at the extreme oblique angles. Just physically rotating the seeker to such angles commonly requires another gimbal axis. Narrower FOV also became acceptable, aiding in IRCM and environmental sources discrimination. The detection range was also extended to beyond WVR, eventually around 6mi (10km) head-on. More sensitive 'prox' fuzes also supported extreme engagement angles and closing velocities. Exploiting these advanced capabilities demanded a highly maneuverable 'agile' AAM strong enough to withstand the very high maneuvering load factors. The ability to maneuver at high AoA and employ body lift can reduce requirements for exceptionally HOBS angles – what are called high-deflection shots in aerial gunnery. Such a scenario is shown in Figures 52 and 53. Such high maneuverability for close-in fighting can still demand very high control authority during the initial period of flight as airspeed builds up. Thrust vector control might be necessary during this acceleration phase. Still, a high-g turn to the side or rear greatly depletes the missiles energy and limits the range for such engagements as depicted in Figure 54. For a head-on pass with a high-speed target, the range to the rear after passing would greatly reduce beam crossing and make rear hits exceedingly unlikely. A very high-speed target may present only fleeting forward quarter shots.

Some of the fifth generation IRH missiles, with extended BVR range (EBVR), also benefit from datalinks supporting LOAL. Trajectory 'shaping', implying lofted flight path, assist in extending range.

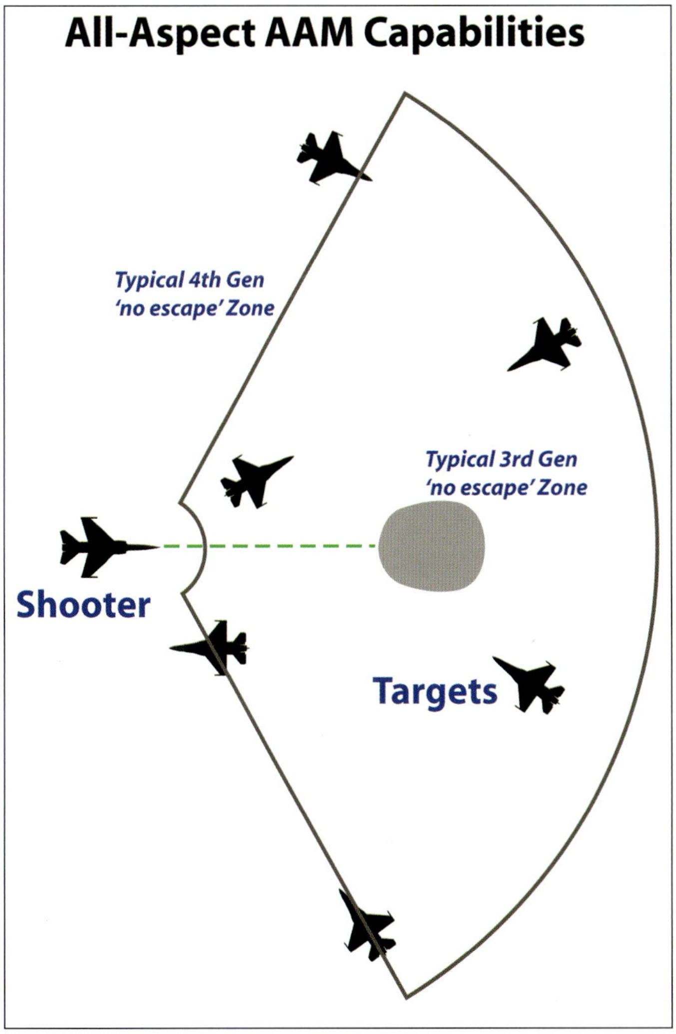

Figure 51: An all-aspect or fourth-generation IRH missile can engage an opposition warplane regardless of that aircraft's heading with respect to the shooter and maneuvering up to the typical limit of 9g. This demands an exceptionally sensitive seeker that was decades before becoming reality. (Drawing by author)

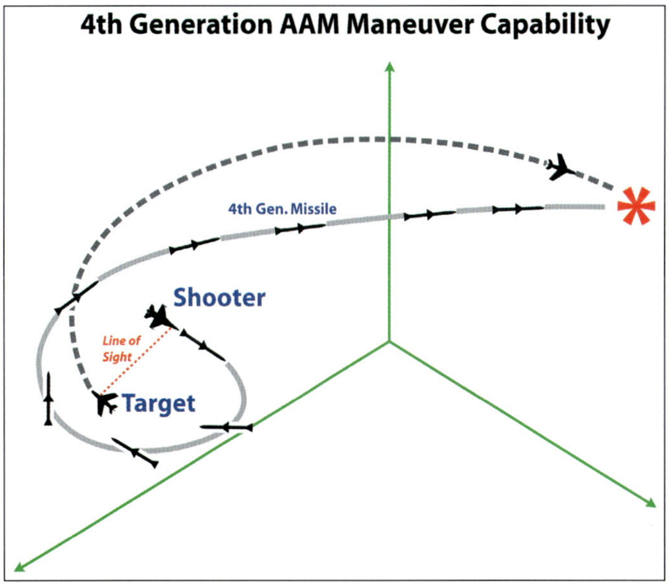

Figure 52: Another feature of the all-aspect IRH missile is the ability to execute a hard turn (very high g and high AoA) and engage in a lead-collision intercept of a target maneuvering at maximum-g. This makes it virtually a 'no-escape' AAM. (Drawing by author)

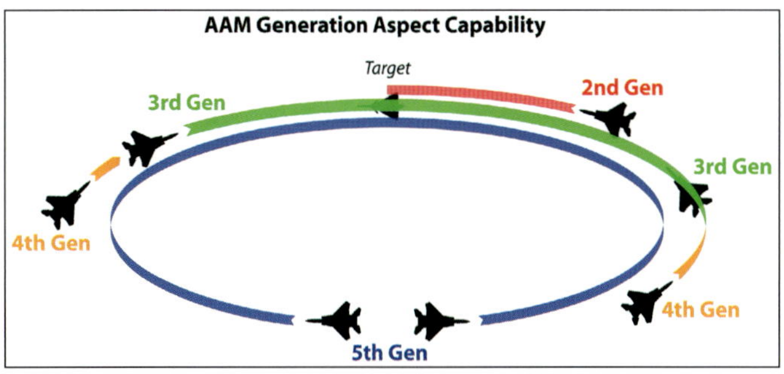

Figure 53: The advantages of the generation of IRH missiles is seen another way in this figure. The second generation (2nd Gen) missile had a narrow cone behind the exhaust of the opponent for effective tracking. The third generation expanded this considerably and permitted some all-aspect capabilities to include forward quarter engagements. The fourth generation expanded the all-aspect capability even farther. The 5th Gen, then, made this absolute to include firing at targets beyond the FOR of the missile seeker. (Drawing by author)

A shot of the South African A-Darter fifth generation IRH missile departing a Gripen fighter (inset) is seen in three sequential frame grabs from flight test camera footage. It is engaging a target well to starboard or even behind the aircraft. Shortly after leaving the rail on the port wingtip the highly agile missile makes an abrupt turn across the flight path of the aircraft on its way to the target, possibly in a lock-on after launch mode. (Denel)

Combinations and Multi-Mode

An ARH missile can have the option to operate in a SARH mode, if so designed. This would guide the AAM semi-actively until the target is acquired by the ARH seeker. The onboard guidance may be designed to make the switch for terminal guidance when S/N favors accuracy of the ARH output. However, the fighter's radar (supporting SARH) and that of the missile (for ARH) may operate in different RF bands, challenging the optimal design of the antenna on the round. More commonly, as previously mentioned, initial guidance via datalink from the fighter (and possibly some off-board source) may guide the AAM to a course where its inherent seeker can detect the target and assume dominance. Such may especially be used by high off-boresight AAM that may not be able to initially detect the target prior to or immediately following launch. This also assists targeting with returns from multiple 'bogies.'

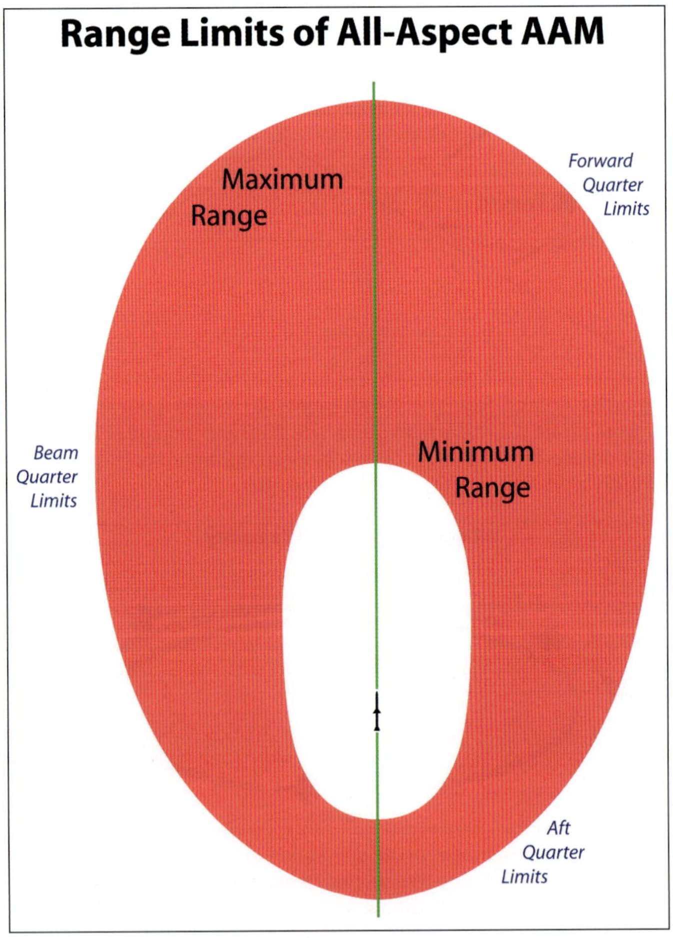

Figure 54: The all-aspect missile is still limited by minimum and maximum engagement ranges. These shrink for beam and rear engagements as energy is bled in high-g turns to engage such targets. (Drawing by author)

There have been a few AAM conceived and tested with a combination IR and SARH to help ensure a hit, especially against low-observable (stealth) aircraft and those employing countermeasures. This, of course, only adds weight and complexity, the latter working against reliability. The IRH elements ahead of the radar antenna introduced boresight errors that must be corrected computationally. However, flying a proportional navigation course, the radar seeker antenna is always off-boresight and so the errors may be naturally minimized if the IRH seeker is installed separate from the radar receive antenna. Figure 55 shows an example design with the IRH seeker affixed to the SARH antenna. The scan rate, FOV and FOR for the IRH and radar functions, individually, may not be compatible and so a compromise adopted. Most likely, neither mode would operate optimally.

More common has been a missile that can have a choice of seeker – usually either IR-guided or radar-guided – installed on a common missile body. This reduces the requirements of accommodating multiple missile types on the aircraft – only multiple models of the same type – while still permitting the aircrew to choose the best weapon given engagement circumstances. However, such has the potential for non-optimal designs with excessive weight from unneeded propellant for an IRH missiles (exceeding the short range of seeker) and too little propellant for a medium-range SARH or long-range ARH missile. Likewise, power generation requirements for these basic types could be markedly different. The American Falcon missiles were the prime example of this trend, with a common AAM body accommodating both IRH and SARH seekers, and the layout scaled up for later types. The USSR created IRH and SARH rounds of the same missile type for many years, offering their pilots tactical options and flexibility while easing the AAM support requirements. In later years this pursuit of commonality was seen as non-optimal in terms of performance, but still found some applications.

A combined SARH and IRH seeker head developed by the US Navy shows an IR seeker at the apex or a radome covering a concave dish antenna, the assembly on a gimbaled mount. Presumably the IR detector is just ahead of the feed horn at the focal point of the parabolic antenna. The IRH can improve endgame intercept accuracy, but design compromises undermine the effectiveness of both modes. (US Navy)

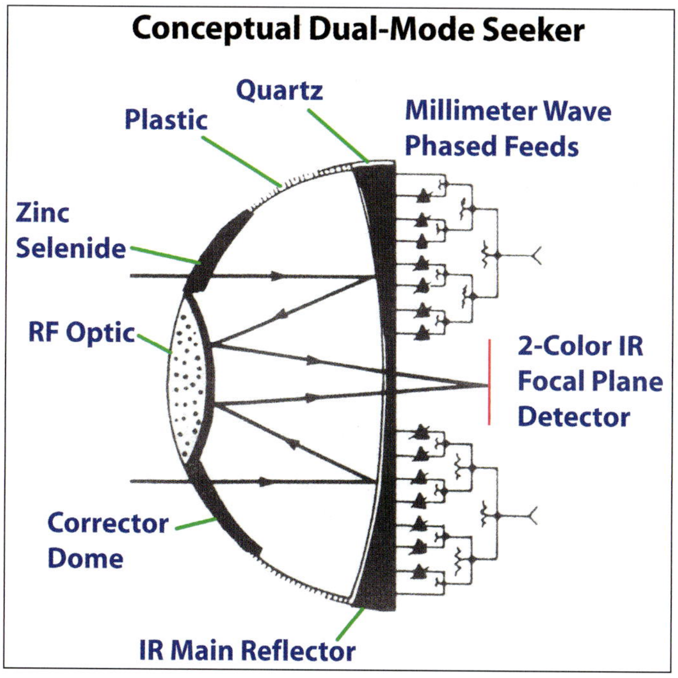

Figure 55: A conceptual dual-mode seeker is shown combining IR and RF (millimeter wave in this case) detectors. (USAF, modified by author)

Right: A good representation of the Falcon series is presented in this image with (left to right) the GAR-11 (AIM-26, nuclear-armed), GAR-2A, GAR-1D, GAR-4A and GAR-3A, a mix of SARH and IRH missiles. (San Diego Air & Space Museum)

Generations of Missiles

To be outlined in the following volumes will be the six generations of AAM, the fifth still emerging and sixth on the horizon. A 'Zeroth' generation represents progress during the Second World War. From the background detailed in this volume, the following generalized characteristics for each generation (Tables 9.1 and 9.2) can be outlined. This is the author's construct as nothing similar has been found in open literature. Each AAM type and model is identified via this summary in a table published in Volume 6.

For the purposes of this work the author has chosen the following range category definitions. Such is subject to much variability in the literature, so this is a generalization. Short range is defined as out to 6mi (10km), medium as out to 15mi (24km), long-range as out to 35mi (55km), and very long-range as 100mi (160km) or more.

Table 9.1: Generalized Characteristics for Generations of Missiles (Zeroth to Third)

Generation	**Zeroth**	**First**	**Second**	**Third**
Period	Second World War	Immediate Post-War	Early Cold War	Mid-Cold War
Development	wartime haste	moderate development/ fielding cycle, modest testing	short development/ fielding cycle, better testing	moderate development/ fielding cycle, greater care in testing including operational trials
Construction	hand-built	hand-built	some machine manufacturing	some automated processes
Electronics	vacuum tube	vacuum tube	transistors introduced	initial integrated circuits
Cueing	none	none	none	initial HMD and radar, cumbersome
Propulsion	mixed rocket	early solid propellant	improved solid propellant	working for reduced smoke
Warhead	blast and case fragmentation	blast and case fragmentation (some pre-fragmented)	shaped charge introduced, some inserted fragments, some nuclear	shaped charge, continuous rod introduced
Fuzing	impact	impact and passive PF	impact and passive PF	impact and passive or radar PF, bias introduced
Accuracy	low	modest	little change	some improvement
CCM	none	none	none intentional	some
Conditions	clear weather, subsonic launch	very modest all-weather, subsonic launch	some all-weather, supersonic launch	improved all-weather, supersonic launch, some internal carriage for drag reduction
Reliability	low	low	moderate	improved
Cost	moderate	low	modest	modest
Longevity	short	short	short	moderate
Short-range	short-range: manually guided via radio or trailing wire, qualified all-aspect	(IRH) narrow-band and uncooled detector, narrow FOR, tail-chase engagement	(IRH) wider-band and cooled detector, improved irdome, wider FOR, higher scan rate, greater angle-off-the-tail engagement	(IRH) cooled wide-band detector, better target discrimination, superior irdome, wide FOV, high scan/ tracking rate, modest all-aspect engagement, longer range, higher-g maneuvering
Medium-range	non-existent	(SARH) cumbersome and poorly performing, qualified all-aspect	(SARH) compatible with continuous wave radar, still modest effectiveness owing to maturing radar technology	(SARH) still limited by radar technology, compatible with pulse-Doppler radar, monopulse, snap-down
Long-range	(ARH) non-existent	(ARH) non-existent	(ARH) non-existent	(ARH) still limited by radar technology
Very Long-range	non-existent	non-existent	non-existent	non-existent

Table 9.2: Generations of Missiles (Fourth to Sixth)

Generation	**Fourth**	**Fifth**	**Sixth**
Period	Late Cold War	Post-Cold War	Twenty-First Century Tensions
Development	long development/fielding cycle, extensively tested including combat-representative operational trials	very long development/fielding cycle, extensively tested including combat-representative operational trials	very long development/fielding cycle, extensively tested including combat-representative operational trials
Construction	considerable machine and automated manufacturing, integrated components, electrical actuators and thermal batteries widely adopted, solid-state gyros introduced	considerable machine and highly automated manufacturing, highly integrated components, solid-state gyros, electrical actuators and thermal batteries common	extensive machine and highly automated manufacturing, highly integrated and federated components, solid-state gyros, electrical actuators and thermal batteries common
Electronics	higher capacity microprocessors, some onboard computing (firmware), built-in test	high-capacity microprocessors, considerable onboard computing (software), adoption of standard data formatting protocols, BIT/self-check, highly integrated with aircraft systems, optimized navigation algorithms (trajectory shaping) for extended range	very high-speed processing, extensive onboard computing (software), possibly artificial intelligence, standard data formatting protocols, BIT/self-check, highly integrated with aircraft systems
Cueing	HMD and radar, improved interoperability, datalink introduced	HMD and radar common, high interoperability, datalink common, LOAL allowing targeting beyond capability of seeker	HMD and radar common, some off-board, high interoperability, datalink common
Propulsion	tailored burning, reduced smoke	tailored burning, reduced smoke, some exotic systems, thrust vectoring becoming common, kinetic range exceeding range of seeker	tailored burning, reduced smoke, some exotic systems, thrust vectoring common
Warhead	shaped charge, continuous rod	shaped charge, continuous rod, aimable introduced	shaped charge, continuous rod, aimable
Fuzing	impact and laser or radar PF, bias more common	impact and passive laser and radar PF, bias common	impact and passive laser or active radar PF, bias common
Accuracy	high, superior target discrimination	very high, very good target discrimination, maneuver prediction	very high, very good target discrimination, maneuver prediction
CCM	very good	very good	excellent
Conditions	all-weather, supersonic launch, improved look-down, some LOAL	all-weather, supersonic launch, some internal carriage for stealth, very good look-down, LOAL common	all-weather, supersonic launch, internal carriage for stealth, excellent look-down, LOAL common
Reliability	good	very good	very good
Cost	high	very high	exceptionally high
Longevity	long	very long	very long
Short-range	(IRH) multi-color seeker, onboard cooling, wide FOR, high scan/tracking rate, HOBS capability, some pseudo-imaging (no reticle), superior all-aspect engagement, some BVR, high maneuverability	(IRH) solid-state IIR (focal plane array), all-aspect engagement common, optimized flightpath, ability to target specific parts of target, super-maneuverable, BVR accepted	(IRH) high-resolution solid-state scene imaging, all-aspect engagement common, ability to target specific parts of target, super-maneuverable, BVR common, effective against cruise missiles and UAV
Medium-range	(SARH & ARH) phased array radar, target hand-off from aircraft radar	(SARH & ARH) highly integrated with aircraft systems, engaging small/low/fast targets (especially cruise missiles) a priority, optimized flightpath, AESA radar sought, SARH fading	(ARH) highly integrated with aircraft systems, effective against cruise missiles and UAV
Long-range	(ARH) more practical, operational experience	(ARH) practical and sought-after	(ARH) practical and sought-after
Very Long-range	non-existent	(ARH) introduction	(ARH) sought-after

4

OPERATION FUNDAMENTALS

Placing an air-to-air missile into service and actually fighting with it requires considerable support and training. Tactics are based on AAM capability, the launching aircraft's capabilities, the nature of the adversary, and the prevalent Rules-of-Engagement (ROE). The probability of a successful engagement is based on a great many factors to include countermeasures (CM) by the adversary. All these aspects will be examined in this chapter. Mission debriefings, with analysis of gun camera film, support reporting from which lessons learned are derived to benefit other aircrew. Military intelligence assessments also benefit from these reports.

Support

The AAM must be taken from stock, assembled, inspected, tested, transported to the aircraft, mounted on the launcher, and connections with the aircraft completed. All this illustrates how supporting an AAM in the field requires considerable equipment like trailers, loading equipment, diagnostic test sets, and the like plus the associated trained personnel and documentation. The desire is to match a new system to existing support equipment to minimize logistics and training impacts with associated costs and delays. Some manner of adjustment to the existing kit for compatibility and identical capability (form-fit-function, F^3) would next be reluctantly tolerated, preferably by adapters or modification. However, just having to change adapters on a loader rig takes time and slows reload in what may be a time-critical combat situation. Undesirable but often necessary is entirely new equipment.

Some missiles may be manually loaded with several personnel physically lifting it up to be attached to the launcher/pylon versus doing this mechanically with a piece of support equipment or a loaded vehicle. This generally means a weapon of no more than 220lb (100kg) and without delicate or sharp projections along the body. This can become a design requirement and drive such compromises as aero surfaces that must be attached only after mounting on the aircraft. This, then, demands as simple an installation as possible with the minimum of tools. Loading, either manually or mechanically, must also be performed in all weather conditions and when personnel are encumbered by special protective gear such as a nuclear, biological and chemical exposure suits.

This film sequence shows the ejection, ignition, and initial moment of powered flight of an AIM-120 from a YF-22 weapons bay. Note the initial nose-low attitude of the missile after departing the bay and the smoke puff at motor ignition. All the equipment, processes, and aircrew actions permitting an effective firing and intercept bespeaks a considerable amount of support and training. (Author's collection)

Top: The loader for an AIM-54 missile was a substantial piece of equipment that had to be moved on aircraft carrier flight deck elevators, and lower decks, plus being supported with its own trained personnel and logistical tail. It had to fit beneath the aircraft and allow fine alignment to mount the AAM on launchers with little tolerance. Banging up the round or the aircraft during this operation was to be avoided at all costs. (US DoD)

Centre: The manually loaded AIM-9 missile had to be light enough for this operation and without sharp edges or delicate protrusions complicating the task. These US Navy personnel are beneath an F/A-18 which had the launcher within arm's reach and the AAM on a transport trailer that fit beneath the aircraft wing. The team is lifting an AIM-9L that has already had the cover for the fuze laser windows removed. The personnel must be careful not to scuff those transparencies. (US DoD)

Bottom: The simplest and most routine tasks are made all the more cumbersome and lengthy by crew encumbered by nuclear, biological and chemical warfare protective garments as shown by these Royal Air Force personnel loading AIM-9Ls on a FGR.2 Phantom. (Jim Rotramel collection)

Common shop test equipment for various types or even models of AAM is desirable but unlikely. Keeping things light, modular, easily operated, and needing only common 'house power' are the usual 'desirements.' Making the gear easily broken down and packaged for transport to a forward operating location is a must and often a demanding design requirement. Installing anything new in this regard on an aircraft carrier can be a major effort involving cutting holes through several decks to lower in large pieces of equipment. That effort requires temporary rerouting of electrical/power wiring, water and fuel pipes, air ducting, etcetera, consuming weeks of effort.

'Standing up' the logistics 'tail' for a weapon is a significant undertaking. This includes stocking unique parts, tools, and equipment. Tracking the stock of spares, anticipating quantities required, and transporting these to operational units are other essential functions. A program office is necessary to track parts usage, submit budget requests for replenishment, then managing contracts with manufacturers, vendors and suppliers. Training maintainers, armorers, and operators requires 'courseware' and specially assembled equipment representing the field gear to include weapons and support equipment. For operators of missiles developed and sold by another nation, support of their weapons might depend upon international agreements to issue alerts of deficiencies and to offer upgrades.

Safety measures are taken in AAM handling and storage to protect against inadvertent ignition. Even without a motor ignition on the ground producing a free-flying projective, the sound, smoke and flame can be exceedingly hazardous. It is likely the missile will slither violently about on the ground and some distance above the surface at very high velocity. An accidental firing within a hangar at Fairey Aviation of a test article during the 1950s had the missile ricocheting about the building damaging equipment, starting fires, and disemboweled an engineer with its sharp wings.[1]

The motor, warhead and guidance section are typically stored and transported separately. The explosive components are stored in a remote, fire-safe 'igloo' with blast doors given that the propellants and warhead can 'cook off' in an intense fire or be ignited by stray electrical charge. These segments are designed to reduce such hazards, in addition to ignition from penetration by an explosion fragment or bullet in combat. Tests are performed to qualify these designs. Even when mounted aboard aircraft, missiles have been accidently ignited and launched, on the ground and in flight, to create an exciting and highly dangerous incident. Accidental warhead explosion has been exceptionally rare.

As one element of the combat load, the weight of an AAM must be known to permit weight-and-balance (W&B) calculations to ensure the aircraft gross weight and CG location remain within limits. For fifth generation fighters these data are entered into the flight computer. Especially as weapons are separated and the combat load changes during a mission, in addition to fuel burn, so does the W&B. This can affect endurance calculations plus handling of the aircraft. The round's weight and CG may also be entered into its guidance computer.

Top: A variant of a USAF Falcon AAM is undergoing checks and maintenance by servicemen using a substantial piece of shop equipment. Such would be challenging to accommodate on an aircraft carrier or transported in a deployment contingency. (Albert Grandolini collection)

Bottom: By contrast, the tester for the AIM-9 Sidewinder guidance and control unit is an easily transportable box with IR imaging source. Presumably there is another test unit for the proximity fuze and warhead sections. (US Navy)

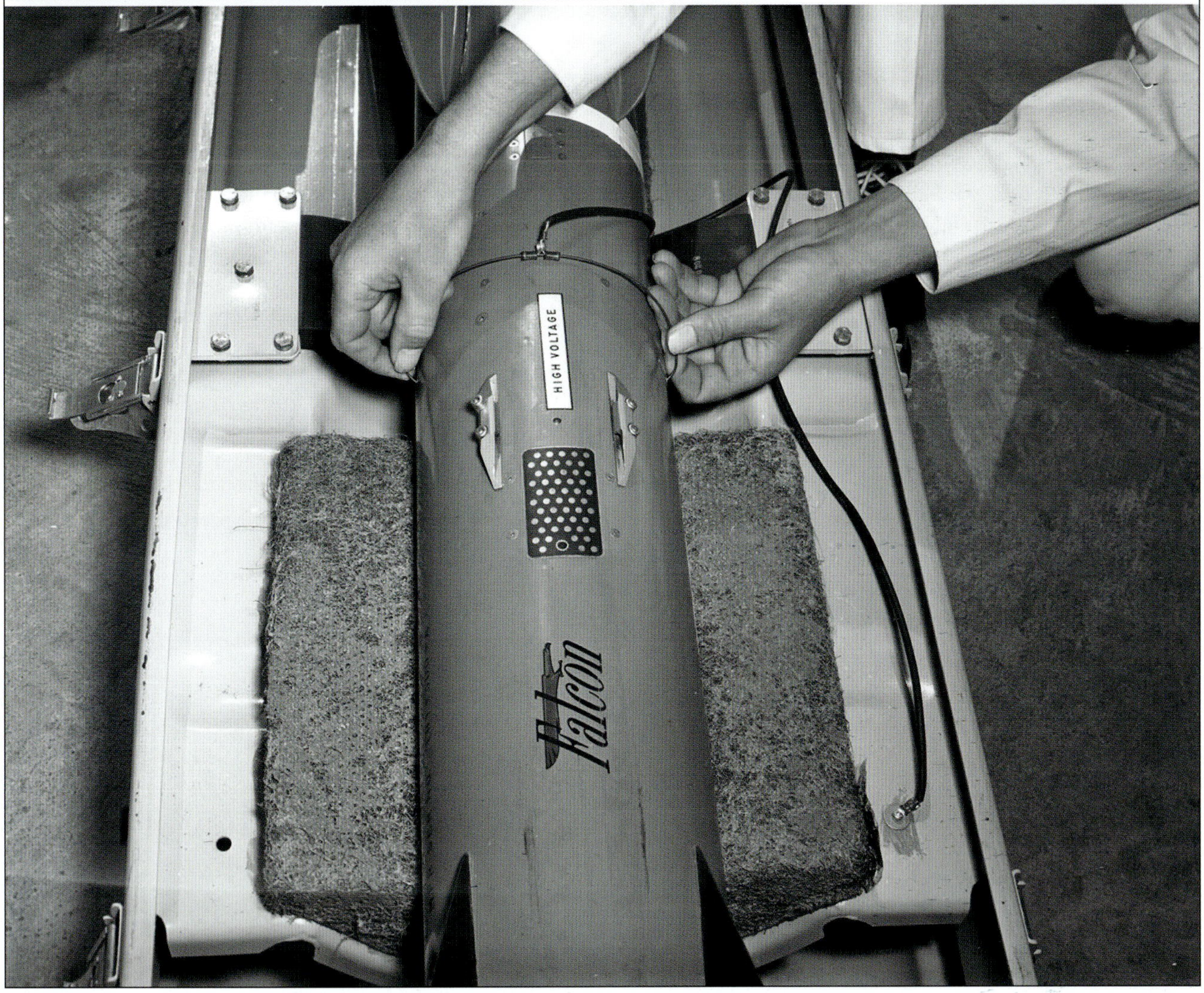

Packing for transport and storage can be vital for safety and preservation of a weapon. Hughes delivered their Falcons in a rugged aluminum and fiberglass case with foam rubber supports and a grounding strap. These weapon loading personnel are lifting an AIM-4C round during an October 1977 demonstration. (Albert Grandolini collection)

Hughes Aircraft technicians conduct a weight-and-balance test of an AIM-120A test article. After a time the weight and CG location of each missile produced in a lot are nearly the same, but the test is conducted regardless to detect deviations and provide data for field personnel loading combat aircraft. (National Naval Aviation Museum)

Like many elements of ordnance, there is commonly some manner of arm/disarm setting to be made by ground crew before the aircraft takes the runway for takeoff. This is done in a safe arming area with the aircraft oriented such that an uncommanded motor ignition and missile fly-out does not hazard other aircraft, personnel, or structures. The initial start and taxi are done in a disarmed state for safety. The arming may be as simple as removal of a safety pin with a 'REMOVE BEFORE FLIGHT' streamer attached that, when inserted into the missile, inhibit the motor igniter and/or arming of the warhead. When removed, these pins are shown to the pilot prior to takeoff for awareness as a double check of the state of the armament. Other safety features such as sensing weight-on-wheels via a 'squat' switch also serve to inhibit arming, firing and possibly jettison on the ground. The pins are reinserted immediately after landing, again in a safety area, prior to taxi to parking and engine shutdown. Other arm/disarm means are rotated switches on the missile, usually actuated with a hand tool.

A missile that has failed to fire (a misfire) or fails to leave the aircraft (hangfire) is considered a 'hung store.' Given that the cause of the failure may not be known immediately, there remains the potential for inadvertent firing or separation owing to a stray electrical charge or sudden jolt like landing. This creates a hazard to the aircraft and those in proximity to it. The aircraft returns to base for a cautionary landing and taxi to the safety location for shutdown and a waiting period of 30 minutes or so. The round may even be downloaded at that site before the aircraft is towed to parking.

Training

Missiles are expensive and difficult to replace, and so firing them in training is rare. To fire a missile against a target also requires some manner of decoy. If these are to be maneuvering targets (typically an air-launched drone) the cost goes up even more. In some cases the decoy is so small and its motor exhaust minimal that it is not representative of a fighter or bomber target. The radar and IR

A US Navy P-2E (P2V-5F) carries two Firebee drones for air-to-air gunnery training or testing of aircraft systems to include AAM. The red paint scheme was common for the targets. (Naval Aviation Archives)

High speed and maneuvering target drones became necessary in the jet era for training and flight testing with AAM. These air-launched BMQ-74C are being completed for delivery to the US Navy at Northrop Corporation's plant at Newbury Park, California, in the 1990s. (Author's collection)

signatures can be enhanced artificially for suitable training. Even a 'boresight shot' against no target, or a shot against a target simulated within the AAM electronics, requires a modification to the missile to provide a false signal of 'lock' so that the pilot can fire the round. Therefore, most aircrew never fire a missile in training and have only lectures, videos and flight simulators of varying fidelity for familiarization.

Missiles can serve training purposes in acquiring and locking-up a target short of firing. To ensure safety, the rounds can be inhibited from firing and/or detonating. In some cases a unique training or flight test round substitutes TM and recording systems in place of the warhead, especially in measuring and recording the closest proximity of the missile to the target if actually fired. The data are used in debriefing the mission for lessons learned and to score simulated 'kills.'

A Python 4 training round is seen at the starboard wingtip launcher of an Israeli F-16I. This supports realistic indications of the IR seeker detecting a target and providing indications to cockpit displays. Consequently, the seeker head remains in place as well as the cooling system as indicated by the panel for refreshing the coolant bottle. The close-out at the end of the airframe suggests that there is no motor and there are also no laser PF windows. All aero surfaces have been removed to make the round easier to handle and to reduce drag for greater aircraft endurance during training. (Israel Air and Space Force)

Man-in-the-loop simulators, such as shown, enhance aircrew readiness for air combat, reducing the potential for confusion and mistakes. Simulators may be fixed base or with motion. They may also be linked with other simulators such that aircrew can fight a simulated battle with an actual opponent versus a computer-generated threat. The behavior of AAM may also be simulated to various levels of fidelity. (Israel Air and Space Force)

User-friendly

Air-to-air missiles are developed by engineers who may have only a conceptual understanding of air combat. It is important that operators, or former operators, be involved from an early stage to ensure requirements are understood, any design compromises are balanced and do not irrevocably undermine combat effectiveness, and that operation is 'user-friendly.' This last is sometimes difficult to explain as many have the impression of fighter pilots as 'cool cucumbers' dispassionately and expertly executing their tasks. Given that most pilots never fire an AAM in training and are very unlikely to have had an AAM fired at them, in actual combat the aircrew is probably experiencing everything for the first time. They may be surprised or confused, reacting in an undesirable fashion such as firing the wrong weapon, firing by mistake or out of limits, or erroneously jettisoning the weapon. Air combat is often chaotic, filled with fear, frantic actions, shouted and garbled communications, and uncertainty. Given coordinating with a wingman and controller, cockpit and weapons management, dealing with a maneuvering opponent, and more, the pilot can easily become task-saturated – especially if inexperienced. The AAM operator may make a mistake under such circumstances, particularly if required to move eyes from outside to inside and refocus on near objects requiring interpretation then action. All this must be designed to be executed as simply, quickly, and effortlessly as possible, and with low probability of error, to increase the chances for success and not create worse circumstances such as shooting down one's own wingman. These factors influence design as guided by operator advice while leaving technical details to the engineers. Design reviews, simulations and operationally representative flight testing help to identify deficiencies in this regard, permitting remedy prior to fleet deployment.

Twenty-first century combat aircraft, with the aid of computers, greatly reduces the efforts of aircrew at system management and permits more focus on tactics and successful engagements. This trend is likely to continue.

Equipment reaction time is an important parameter for the pilot in air combat as well as understanding how the systems work. For example, the selection of missile on an armament panel or touch screen, gunsight or HUD adjustments, arming and firing sequence (Figure 56), must all be as simple and easily interpreted as practical. Any automated features must also be quick, readily comprehended and evident to the pilot. Half a second delay can mean the difference between victory and failure. With many pilots firing their first AAM only in combat, it becomes essential that they know what to expect and not be so startled that they lose situational awareness. An example of poor design in this regard was changing the US Air Force's F-105's gunsight from the air-to-ground to the air-to-air mode. This required five separate switch movements, some of these switches at difficult locations in the cockpit. Performing this task accurately while keeping eyes on the threat and maneuvering hard was exceptionally difficult. Consequently, during Vietnam combat the pilots often chose to fire the cannon without benefit of the lead-computing gunsight. Selecting the AAM was five additional switch movements that had to be performed in the correct sequence.

This F-4E illustrates the 'office' of a fighter pilot and, in the Phantom's case, the radar and weapon systems operator 'guy in back.' The aircrew are trussed-up in anti-g suits with air supply, possibly anti-exposure suit and floatation devices, a survival vest, maybe a sidearm if going into combat, a helmet with earphones and visor, and mask with microphone and a hose to the oxygen supply. They are secured to an ejection seat and its hard cushion and four-point harness. They are surrounded by instrument panels and controls. The pilot has a head-up display combiner glass ahead. Vision is limited by canopy rails and bows, and portions of the aircraft. The cockpit can be blistering hot or frigid cold despite the air conditioning. It is fair to say this all takes some getting used to if trying to fight successfully. (USAF)

The complexity of the cockpit in a modern fighter aircraft is suggested by this image of an F-16I with all displays and indicators illuminated. Some controls may seldom be used while others may have multiple functions that the aircrew must memorize. Successful air combat demands proficiency in managing the aircraft and the weapon systems. This can be a high workload at times, requiring continuous training. (Lockheed Martin)

A personal account of how bad a miscalculation can be, from the perspective of a fighter pilot, was related by US Air Force Colonel Robin Olds, leading a fighter outfit in Vietnam. He had a strong reaction when a technical representative ('tech rep') came to introduce his crews to the new heat-seeker AIM-4D Falcon missile being fitted to their F-4D Phantom IIs jets. These were displacing F-4Cs that had been equipped with Sidewinder AAM. After a skeptical Olds asked that the basic description be repeated twice, he erupted:

> "I don't give a rat's ass where or by whom it's been tested and accepted. I'm telling you the switchology you describe doesn't fit our situation. Whoever thought up that gadget didn't know a damned thing about air-to-air fighting, and didn't ask anyone who ever had."
> I left the briefing room in anger, not at the tech rep, who was just doing his job, but at whatever bunch of designers, developers, and producers had put that creation on an air force fighter. I wondered what command wanted the damned thing in the first place. The Falcon required a complex set of steps to cool the seeker before the missile could be fired, sometimes taking up to a minute. Once cooled it was good for a limited period of time and couldn't be cooled again. This was intolerable in dogfights. Who in the hell thought we had the luxury of checking our watches? Unlike the Sidewinder that had a proximity fuze and a sophisticated warhead, the AIM-4 had a contact fuse, which required a direct hit to explode. Assuming it locked on, launched, and guided, it had to hit the target with its little 3-pound warhead in order to do its job. What a classic fuck-up.[2]

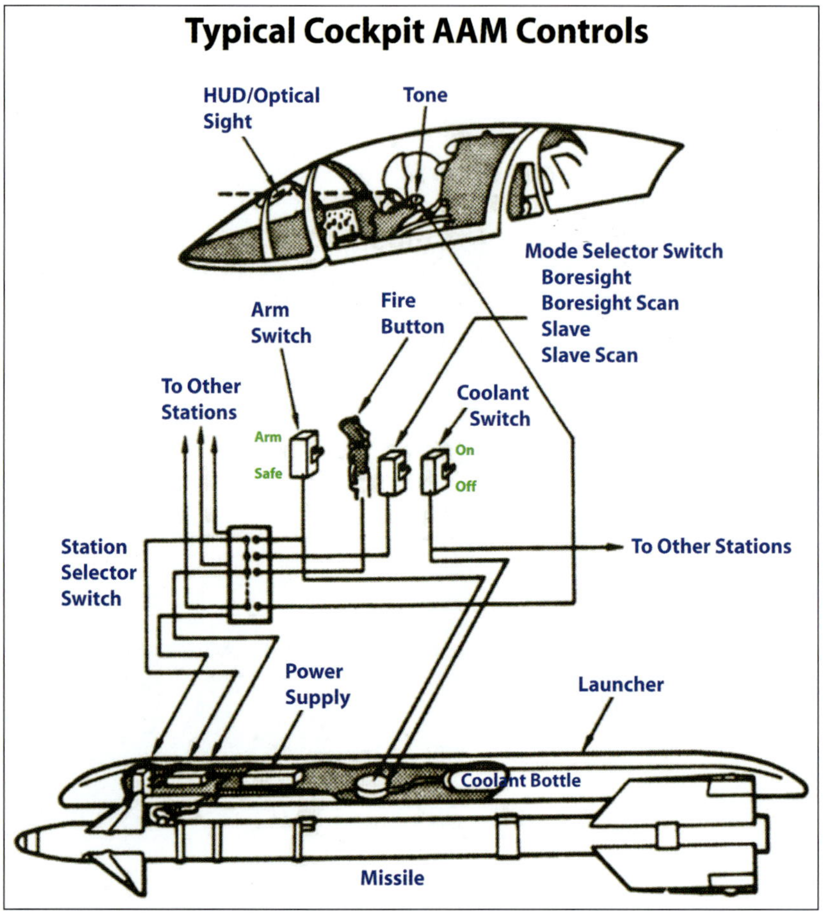

Figure 56: The cockpit controls for the AIM-9L launcher/missile is depicted along with the helmet earphone tone indicator. This serves as an example of such controls and AAM interface for other systems. Some are more complex, and switches may be virtual selections on a touch-sensitive screen in the instrument panel. (USAF, modified by author)

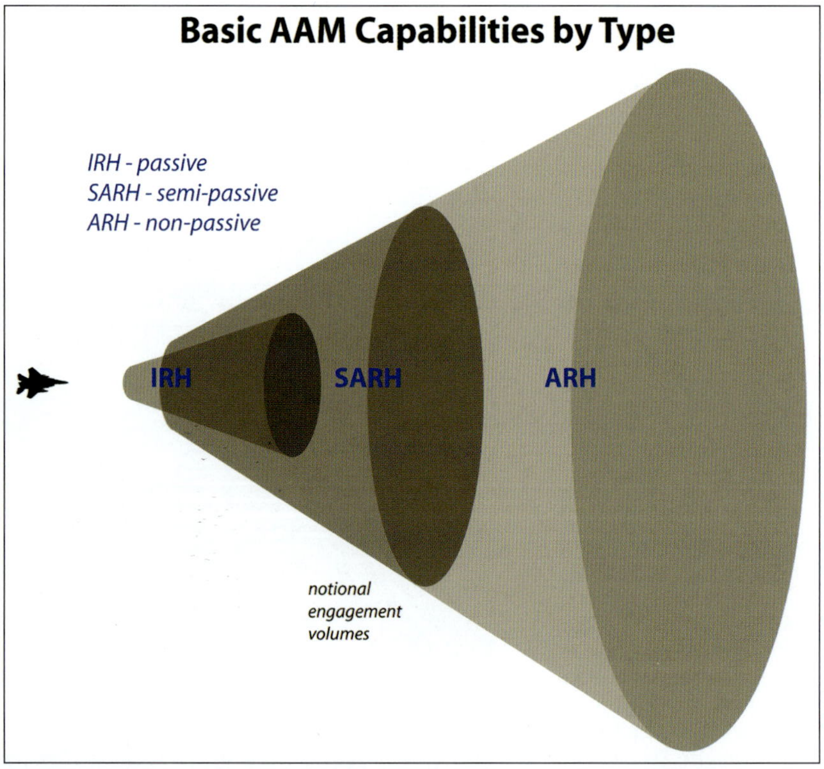

Figure 57: The engagement scenario and systems available (weapons onboard and systems' status) will dictate, to a large extent, the AAM to bring to bear. Weapon capabilities such as FOR, range and autonomy are vital in this assessment. This drawing shows simply the minimum and maximum range of the three basic types of AAM and engagement cones ahead of the shooter based on either the seeker FOR or that of the aircraft radar. (Drawing by author)

Intercept

Fundamentally the fighter aircrew must locate a target, identify it if necessary, select the best weapon to employ (Figure 57), close to the point the AAM seeker attains a lock (unless fired in LOAL mode), then fire the round. All these steps encompass many onboard and off-board systems plus aircrew judgment in a complex chain requiring continuing training to be performed quickly and successfully. The target aircrew and supporting assets will work to prevent the 'kill' to include employing CM. The side with the superior technological assets may prevail, but the extent of personnel proficiency with the equipment and determination can undermine such advantages.

Battle Management

Fighter aircraft aircrew finding their target in the vast expanse of the sky is a perennial problem. By the advent of AAM the vectoring of interceptors via radar detection had become common. This could begin with radio instructions from a surface or airborne radar station, but closing on the target is

Ground control of the air battle may depend on the circumstances, equipment and personnel. This may be a well-established 'brick and mortar' facility, or a mobile unit as suggested by this image of a USA military air traffic control unit in Afghnistan, likely in a trailer. The commander for directing engagements would most often be an officer. (US DoD)

typically via onboard radar unless seeking to remain undetected by adversary RWR.

Ground-controlled interception (GCI) is the common means of vectoring a fighter to a bogie observed by surveillance radar. This is a radar ground station with personnel monitoring the airspace to manage traffic and detect unauthorized flights. The best-placed fighter aircrew are given bearing, range and altitude (BRA) data on the target to effect the intercept. The AEW and AWACS aircraft, with radar and controllers onboard, can detect targets and direct an intercept as well, possibly even identifying the adversary aircraft type based on their point of origin (squadrons at a specific air base) and flight behavior. Being airborne, the radar can sense farther beyond the curvature of the Earth than the ground station. A less costly alternative is a tethered aerostat carrying aloft a surveillance radar. All rely upon 'skin-painting' of airborne radar echoes as well as Identification Friend-or-Foe (IFF, a radio transponder) to identify and track aircraft. 'Friendlies' would typically 'squawk' an assigned IFF code whereas 'hostiles' would not. If intelligence agencies have provided suitable frequency and code data, adversary IFF may be monitored. The IFF has not always performed reliably, and 'spoofing' of the signal by the opposition can complicate its use in tactical decisions. Managing the air battle still requires an officer prioritizing targets and vectoring suitable resources for intercepts. The present-day goal is the merging of data from many sources, to include satellites, the AWACS, aerostat, ground stations and radar of other fighters in the airspace, all passed via datalink to a central processor, for a more complete picture of the battlespace.

It is important, especially with 'all-aspect' missiles possessing a fire-and-forget capability, to avoid scenarios in which the missile locks on the shooter's wingman or other friendlies. Air battles are commonly executed under ROE that dictate how the aircrew are to prosecute engagements. These may require visual identification of targets before shooting, and so disregarding the range of radar and missiles permitting BVR firing. The aircrew may then be required to await verbal authorization to fire. Indeed, one US Marine Corps F/A-18 fighter pilot instructor relayed that only five BVR kills in combat had been recorded to May 1991.[3] This became especially true after the 1970s when – aside from an all-out war – shooting down aircraft came to have significant geopolitical implications. Consequently, BVR capability for infrared guided AAM has not always been pursued even when technologically achievable.

Targeting

Likely after vectoring by GCI, early detection and identification of a target is usually achieved by onboard systems with capabilities superior to the missile seeker. These may include radar, non-emitting/passive infrared search and track (IRST) and targeting pod optics like television – all of these with varying capabilities to detect, identify, track, and cue – plus calls from other aircraft whose crew have made visual contact. All rely on the aircrews' 'Mk 1 eyeball' for operation and identification of the target to the extent the systems permit. Fully automated engagements without human intervention have only rarely been fully embraced because of the potential for fatal errors. Ultimately, it comes to the target being designated on the HUD or

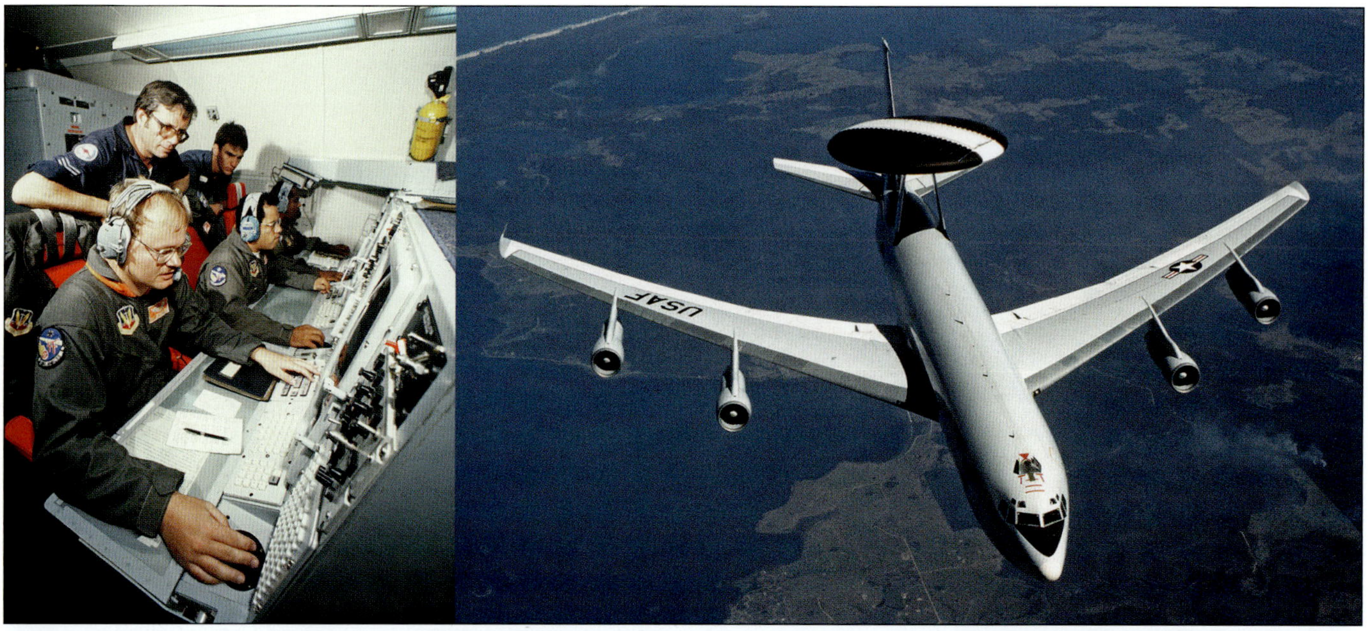

Airborne warning and control assets are expensive and so deployed only by the wealthiest nations or as part of regional alliances. These are typically large aircraft with accommodations for powerful surveillance radar, operators and air traffic controllers. This US Air Force E-3A Sentry is a fine example, the 'rotodome' on top rotating to affect radar scans of the antenna inside. Interior stations include the controllers and commanders who can prosecute intercepts. (US DoD)

HMD for engagement, the latter also permitting cueing. Aside from radar and possible laser rangefinders, the passive means lack range data save for possible estimates if the aircrew actually see the target. Lasers for ranging air-to-air prior to an AAM shot is not known to have been practically applied, given the limitations of power and light scattering in the atmosphere, but has the potential for a passive means of obtaining the data. Onboard radar is still vital in tracking adversary aircraft in the airspace, with range, speed and altitude, to develop the battle to advantage in closing for the engagement. Even if it develops into a close-in fight with short-range AAM, the radar indications on a HUD directing the aircrew where to look for the target can aid in the earliest visual detection. All these means build confidence in a potential engagement while closing to the point that the AAM detects the target. If so-designed, cueing can permit AAM firing, without aircraft maneuvering and seeker acquisition, in a LOAL mode. In this case, the missile initially steers to a commanded bearing or heading and elevation until the seeker detects the target and assumes its own guidance. Additional aircraft input to the AAM guidance during flight via a datalink is also relevant for some products. Notionally, this permits shooting at targets outside LOS and even behind the launch aircraft. Some fifth-generation systems have also been designed to obtain these data via an off-board source, such as an AWACS with radar capabilities exceeding those of the shooter, the data uplinked to the fighter for download to the missile. Such a 'hand-off' can shorten the detect/lock-on/shoot cycle.

The importance of early identification is to permit engagement at the farthest range allowed by the AAM system and permitted by ROE. A BVR engagement negates the need to close for visual identification, which may be inside the minimum range of the system, and then maneuvering to open the distance before firing. The IRST is an IR-band EO telescope that the aircrew can use to scan the area within the system's gimbal limits to detect IR emitters. It is like the seeker in an IRH missile, but much more sensitive and capable. It presents an image on a cockpit display of the target, if properly sighted, for identification plus bearing and elevation from the aircraft. The use of IR versus visual light is to permit this identification even at night. Visual light televisions have also been employed in targeting pods commonly used with air-to-ground weapons delivery. Past IRSTs were a low-resolution television picture whereas later units used focal plane arrays and computer displays for greater resolution and no television-like scan bars. Similarly, visible light targeting pods moved from TV to charge-coupled devises (CCD) for a more refined image. Range data are not available from either of these systems unless coupled with a laser rangefinder that has its own limitations and has not been employed with AAM to the author's knowledge. The simplest early sighting systems were used as a means of turning the radar to detect a target seen on the IRST or TV display to hopefully attain a radar lock. The utility was marginal and, with advances in radar, the IRST disappeared for a time. With the need to identify targets before engaging, the IRST reemerged but with the capability to display an IR scene at various magnifications. It can track the target and the data used to cue an AAM seeker.

Engagement Scenario

Fundamentally an intercept involves range and aspect angle between the shooter and target with consideration of the AAM likely acquiring then tracking the target. Altitude and look-up or look-down plays into this last. The choice of AAM to employ, assuming more than one, must be weighed by the aircrew and the decision potentially made quickly. The selection depends on the nature of threat (particularly signature), range (minimum and maximum), adversary maneuvering or maneuver capability, adversary countermeasures available or being employed, available CCM, plus urgency of a 'certain' kill. 'Speed is life' as much in AAM battles as other aspects of air combat. Shooting a very fast missile that can close rapidly on the adversary before that aircrew can fire its own AAM may be critical to success and survival (closely linked). Tactics for a successful engagement, then, demands evaluation of the most effective exploitation of the warplane and missile combination. Prior preparation, knowing the combat area threats and likely adversaries (derived from intelligence assessments), can help a commander assign the most advantageous choice of these assets plus aircrew to dominate the air battle. Additionally, AAA and SAM threats can distract from the intercept task and must be dealt with or avoided.

The chin pod just behind the nose radome on the US Navy F-14 (left) has a Television Camera System (TCS) optical head on the left, a TV telescope. On the right is an infrared search and track (IRST) optical head. Together, these enhanced target identification beyond that possible with the naked eye and permitted engagement with more confidence that the target was hostile. Such passive systems have been used on several fighter types to provide targeting data to the fire control system and occasionally slaving to the radar. A more common IRST installation is shown at right on a Su-27. (US DoD and Public Domain)

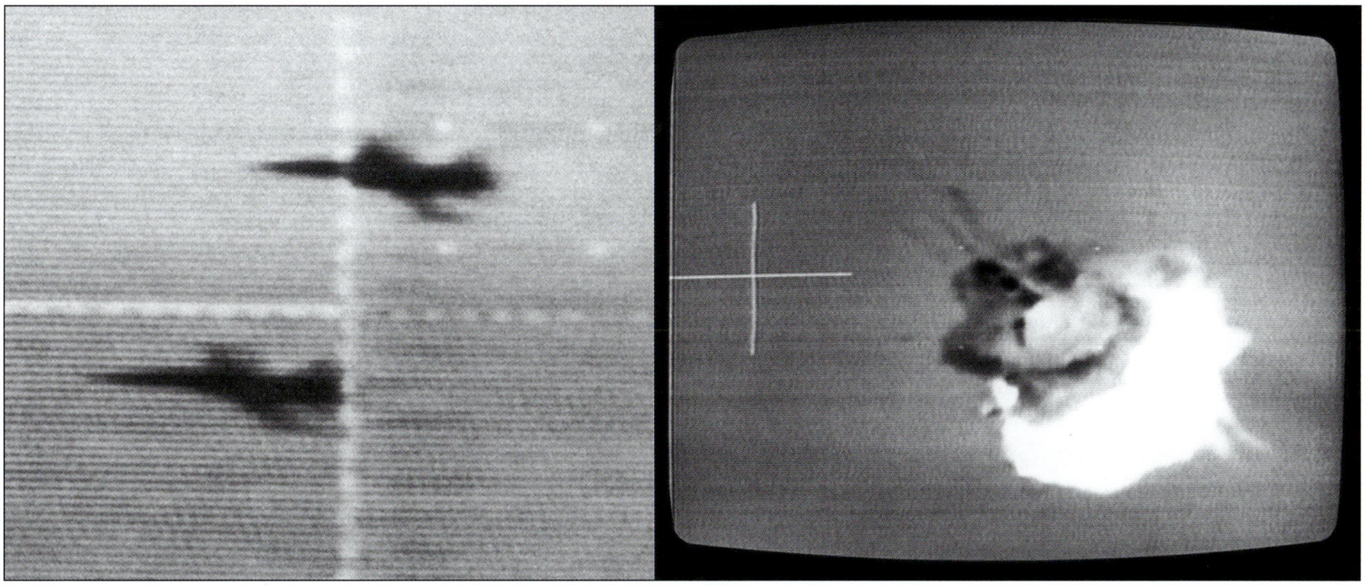

At left an image from an F-14 Television Camera System (visible light) shows a flight of F-5Es at 12mi (19km) range. Would this be sufficient for positive identification before firing a beyond visual range missile? This is likely a zoom-in on a portion of the television screen as the image at right shows a Phoenix missile kill of a QF-4 drone by direct hit (no warhead) recorded from the same cockpit display. (US Navy)

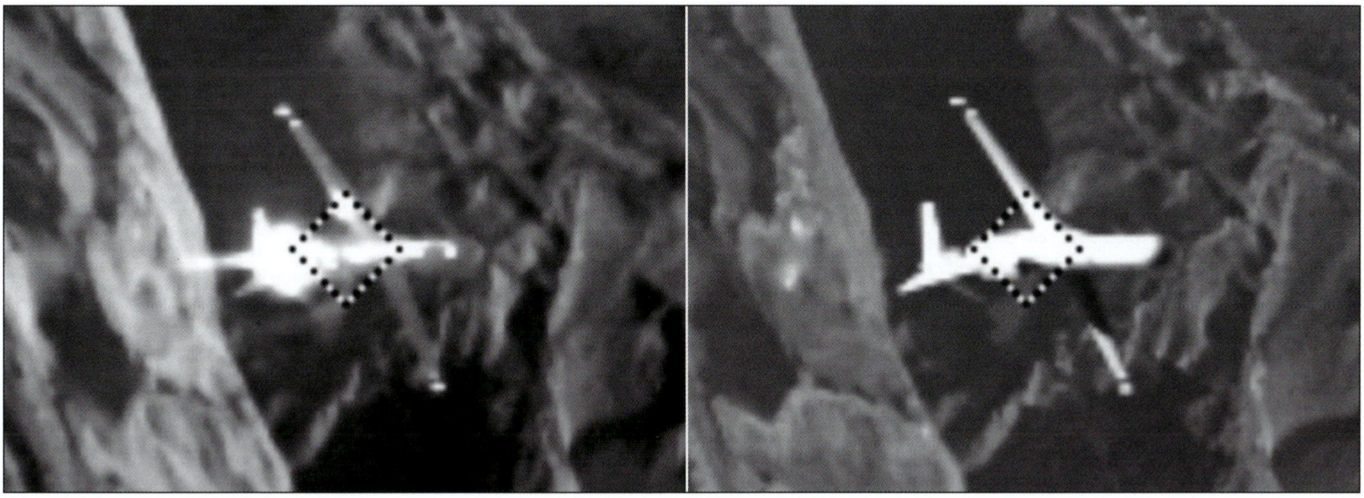

Two targeting images from two different devices give an impression of the identification task facing aircrew in the current era. At left is a picture from an imaging infrared system, likely a modern IRST. In the IR spectrum the UAV is still relatively easily identifiable, at least on the scale it fills the frame, though the glare from the engine and exhaust is distracting. At right is the same picture from a visible light charge-coupled device (CCD), so also a modern solid-state system. (Rafael)

When it comes to the actual engagement, the target is tracked, and the shooter closes to firing range. Depending on the range, target maneuvering, and type of AAM employed, this may require some maneuvering on the part of the shooter to gain the best conditions (range, aspect, background) for the greatest potential for a successful shot. If the target has a performance advantage, the shooter must compensate by employing the best combination of own aircraft performance and missile capabilities – tactics again. The target must be acquired by the AAM sensor and then indicate it has a lock, ready to be fired. This is typically an aural indication, a tone in the helmet headphones, but may include a visual as well such as a symbol in the HUD and/or HMD. The aircrew will have prepared the aircraft via the MASTER ARM switch, selected the weapon, and pulled the trigger on stick or throttle to launch the AAM. There may be an intermediate step; a trigger pull to a first detent for a final indication of ready to fire and arming of the launch circuit, then pull to a second detent to launch. Once the missile is away the aircrew may need to turn away to ensure against being engaged by another adversary or keep a radar lock on the target depending on the nature of the AAM.

Combat may begin with a head-on pass permitting positive identification of the target as hostile. If choosing to engage with a missile, the fighter pilot must then maneuver the aircraft to permit the seeker head to acquire the target. This may be minimized or even eliminated to some extent by the capabilities of the missile, even permitting an immediate launch when certain the target is hostile. This would require an AAM with super maneuverability, cueing and LOAL, qualities only available in late-generation types.

Firing a WVR missile like an IRH round at near maximum range, provided there is a solid lock, eliminates the need to maneuver for a shot solution. The limitations of early AAM – IRH against IR reflections and radar-guided with ground clutter – could mean setting up attacks from below the target with the sky as a background. The AAM also has more time to deal with maneuvering targets with a solid guidance solution and fewer energy-draining sharp maneuvers. Additionally, the motor may well have burned out by the time the targeted pilot seeks to make a last-second evasive maneuver and so, lacking a smoke trail to observe, is unaware of the AAM's proximity. Driving in close for what might be perceived as a surer kill also raises the potential for enemy aircraft to turn and more readily engage in a dogfight that may not end well for the shooter. The aircrew decision of when to switch from radar scan to track mode may also be critical if the adversary, detecting this via RWR, takes evasive action and employs CM early enough to spoil the BVR attack.

The opportunity to get off the first AAM shot can be decisive. Especially the ability to shoot first with a BVR missile, when within range, can be crucial in the outcome of the air battle. This is especially true if it has the potential for destroying the radar platform for a SARH missile about to be fired by the opponent. Before closing, the pilot may choose to fire long-range radar-guided missiles to 'thin' the opposing force. Even if the AAM misses, having a missile pass through their ranks can 'spook' the opposing aircrew and cause them to break formation, losing the benefit of mutual defense and simplify radar homing AAM targeting. Consequently, some pilots may 'fire-for-effect' rather than with confidence of a hit. The effect may create a more favorable engagement scenario. However, firing on a multi-bogey formation means the radar-guided missile may home initially on the center of the mass reflection and only guide to one aircraft late in its flight, potentially losing energy.

Missiles of different types may be fired sequentially in anticipation of favorable time/range-dependent circumstances developing or avoiding the potential effects of the opponent employing countermeasures. This may be seen as a radar homing AAM launched followed by a heat-seeker as the firing jet drives in towards the target. However, the latter round may home on the rocket plume of the former, and so the order of firing can be important. The early missiles using bang-bang control design experienced an erratic flight. Some pilots, observing this, assumed the missile had malfunctioned and launched a second before seeing the results of the first shot. Also, the low reliability of the early missiles meant crews were trained to fire more than one round at the same target to 'raise the odds.'

Whether BVR or WVR can determine the choice of AAM to employ, assuming the aircraft is so loaded. The nature of the target is another factor in this selection as well as flight conditions and knowledge of available countermeasures. Likewise, closing to effective range without being detected can also be a critical factor, especially if the target is not maneuvering or is maneuvering leisurely. Historically 70 percent of kills were scored against an opponent unaware they are under attack and flying at less than 30deg angle of bank. The engaging pilot must close to within the range of the AAM as judged by target size within the optical gunsight/HUD (Figure 58) or as indicated by the ranging function of the radar. It is possible to be too close given minimum range – the time/distance required for the missile to perform functions such as uncage seeker heads, warhead fuzing delays for safety, reaching an airspeed enabling sharp maneuvering, and executing guidance/flightpath correction actions. There is also concern with bits thrown off the target, after warhead detonation, impacting the shooter aircraft unavoidably passing through the debris cloud. Generally, 1,000ft (300m) has been the practical minimum engagement limit in jet combat. In circumstances where the pilots may have closed to gunfire range and then has to resort to missiles, it may be necessary to open the distance before firing the AAM.

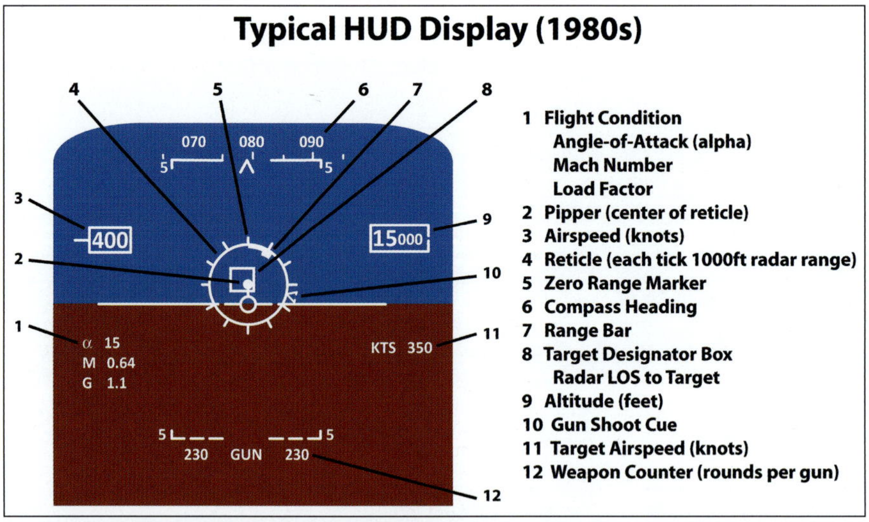

Figure 58: A typical HUD display circa 1980s is shown. Flying the aircraft such as to place the pipper on the target, or within the target designator box, may be a pilot task prior to launching an AAM. Visual indication of weapon readiness and lock are other indicators one could expect on a HUD. Later wide-angle HUDs had more 'real estate' for such additional information, though it is possible to overload the pilot with data. (Drawing by author)

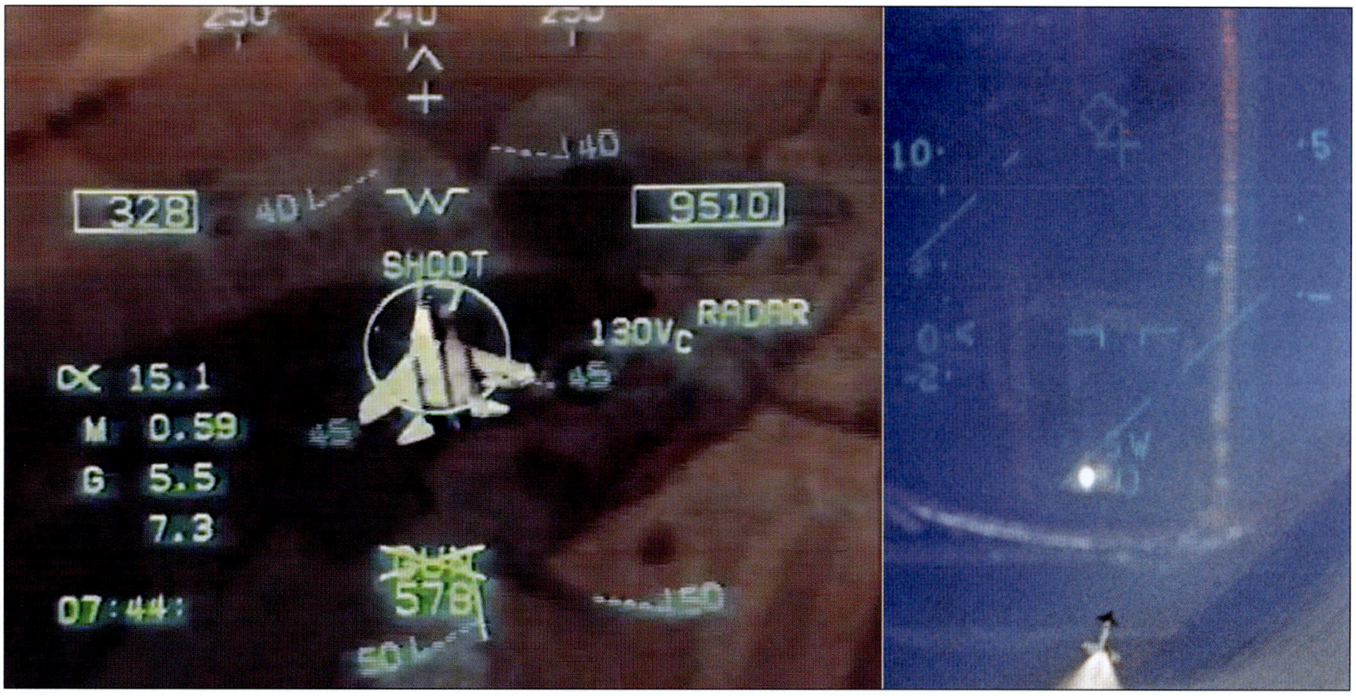

Two frames of gun camera film, with HUD symbology give a good pilot perspective of an engagement. The one at left is from an F/A-18C during air combat maneuver training against a German MiG-29. The one on the right, shows a Sidewinder streaking away from the turning F-14A. The AAM leaves a smoky trail and heads towards a target near the crosshairs (orange dot, upper center). (The bright dot above the missile is most probably a reflection.) The objective was an MQM-74 drone, evidently maneuvering, over the Pacific Missile Range in 1985. (US Navy and National Naval Aviation Museum)

Care must also be taken when in a multi-aircraft engagement to ensure an AAM that misses its intended target does not inadvertently lock onto a 'friendly.' A good example is a fighter attempting to shoot down an opponent chasing a friendly. Firing an IRH from a rear aspect may be a good choice, but if the opponent evades the AAM the seeker may then lock onto the squadron mate previously under attack. It is less likely that a radar-guided AAM locks onto another aircraft. It is not inconceivable for a SARH missile to home on an aircraft that happens into the very narrow radar signal, or even ends up seeking the source aircraft emitter, but these would be extraordinary incidents.[4] Another scenario is a wayward missile happening to pass close enough to an untargeted aircraft for the PF to set off the warhead. Even that would be exceptionally remarkable, the aircrew suffering the hit undoubtedly among history's unluckiest aviator(s).

Knowing when to break off an engagement can also be critical. Fuel state is one factor in this timing. Fighters burn fuel at a prodigious rate when in combat with many rapid throttle transients (some into afterburner) and changes in altitude. (Any external tanks are commonly dropped before engaging.) Mission duration can be reduced to less than 30 minutes in such circumstances. Consequently, aircrew are taught to avoid having their attention overly channeled and to check fuel remaining often enough to avoid reaching the point they cannot return to base or reach a tanker track. There is a 'joker' quantity at which the aircrew have options to divert to an alternate airfield, if necessary, and 'bingo' where a safe return to the primary base is achievable. Below bingo the aircrew will likely be declaring a fuel emergency and potentially landing 'on fumes' at the nearest runway. These numbers are determined prior to the flight and kept uppermost in the minds of the crew. Some fights have ended with aircraft reaching fuel exhaustion and engine 'flameout', then attempting to land on a road and coming to a bad end. This can count as a 'kill' for the opponent. Aerial tankers can be vital at ensuring recovery of an aircraft at a critical fuel state and also to extend fighter patrol time such as to make the most of available resources. However, only the wealthiest countries have been able to afford tankers, most in few numbers, and some fighters are not equipped to refuel from a tanker. 'Bugging out' may also be necessary if the adversary has been joined by companions such that numbers are no longer favorable. Being out of missiles and cannon shells will force the issue as well. Finally, the commander supervising the battlespace can decide it is time to disengage based on factors unassociated with the individual fight.

It may also be wise for the shooter pilot to slow the aircraft after firing, and even turn away if practical, to reduce any advantages the targeted pilot has in immediate reactions. One such is a SARH missile fired at medium range in a head-on engagement, but the targeted aircrew launching an IRH missile before the SARH arrives, as the aircraft approach each other to close range. The heat-seeker may still hit the medium-range shooter after the launch jet has been destroyed. There have certainly been situations with opposing missiles passing each other towards their targets. The lethality of the missile against the target is similarly vital as there may be just one opportunity for the kill.

Aspect Angles

Figure 2 depicted the geometric relationship between the attacking aircraft and the target aircraft. The aspect angle was defined as the angle between the pilot's (or missile seeker's) LOS and the target's flightpath. This is also associated with the AAM's crossing speed in meeting the target, the greatest being head-on and next at an aspect angle of 90deg (Figure 7). For a boresight shot, then, the target must be directly in front of the shooter, though this is not necessarily advantageous especially if crossing (very briefly). Except when the target's crossing velocity is zero (directly head-on or tail-on), some lead angle is necessary when firing a gun for the round to meet the target given the relative velocities and time of flight for the round. Missile guidance systems must also perform a lead calculation for an efficient flightpath, though lead angle is limited by the FOV or gimbal stops of the seeker for maintaining LOS on the target. If the target is

Aerial refueling can extend flight duration for extended patrol and other combat scenarios. This is referred to as a 'force multiplier' as it can amplify the combat potential of existing air assets. With such resources an interceptor can remain on station for hours beyond the usual fuel exhaustion point and has more potential to bring weapons like AAM to bear on the enemy. Shown is an Israeli Boeing 707 airliner modified with two hose-and-drogue pods near the wingtips to refuel probe-equipped fighters and a tail boom to refuel receptacle-equipped fighters. (Author's collection)

Few aircraft have the capacity to carry and deploy more than two types of AAM, typically SARH and IRH. The powerful and complex US Navy Tomcat was able to carry AAM filling the short-range, medium-range and long-range regimes as shown by this F-14A+ with (foreground to background) AIM-9M, AIM-7M and AIM-54C mounted just in the visible portion of the aircraft. (USA DoD)

performing evasive maneuvers then the aspect angle is changing – and so is crossing velocity – at a rate associated with the rate of turn or lateral acceleration measured as g. Maneuvering by the shooter contributes by altering the LOS and even boresight on the target owing to factors like sideslip and AoA plus airframe flexure. (A description of fighter maneuvers to gain favorable shot conditions is beyond the scope of this text.) If the pilot shoots with LOS behind the target (lag) the missile has to turn sharper to establish a lead angle and experiences greater LOS rates that could drain energy and even cause loss of lock. It may then enter a lag pursuit, overshooting the intercept behind the target and then entering an inefficient tail chase (Figure 24). The range to the target is likewise changing for all the pursuit curves in the figure. Clearly, the lead-collision trajectory is the ideal.

If the targeted pilot executes an evasive maneuver, the missile will attempt to follow. Given the typically much higher speed of the missile and so closing rate, attempting to react to the LOS rate change to complete a lead-collision intercept will demand even more g than the target aircraft – a rule of five times is promulgated. The late-generation high off-boresight capable and all-aspect missiles have typically been capable of maneuvering up to around 35g, demanding a very stout airframe.

Figure 59 depicts the basic engagement envelope for the first few generations of IRH missile; essentially tail engagements with small off-boresight dependent on the capabilities of the seeker and any exhaust masking by the target aircraft's design. With the most intense IR energy from the turbine, afterburner flame holders, and adjacent casing, these elements will be masked at roughly 30deg off centerline. The seeker FOR may not permit detection to even that angle. The meaning of the minimum and maximum range was presented in Chapter 3. This is against a non-maneuvering target. A turn would slew the effective engagement cone behind the target jet and greatly limits the shot opportunity as shown in Figures 60 and 61.[5]

Figures 62 and 63 show the top views of the conceptual engagement envelops for an AAM against a non-maneuvering and a maneuvering target and can be conceptually expanded to three dimensions.[6] Maximum range to bring the AAM to within warhead lethal radius is defined as the limits of missile performance (kinetic and seeker), shooter and target airspeed, and target aspect. The head-on and forward quarter shots clearly have greater potential for 'hitting' the non-maneuvering target given higher closing speed compared with tail-chase and rear quarter engagements (Figure 7). The seeker limit is the distance at which the seeker can detect and solidly track the target – a function of many factors as presented previously in Chapter 3 – though the AAM can be fired beyond this limit if in a LOAL mode. For the non-maneuvering case shown in Figure 62, the kinematics of the scenario exceed the capability of the seeker in the forward quarter but is beyond the maximum range line elsewhere. Minimum range zones are, as described in Chapter 3, determined by such factors as guidance reaction time, time to reach airspeed supporting high control surface effectiveness, AAM turning capability, seeker uncage time, and time required for fuze arming safety margin (higher closing speeds amplifying this). As mentioned previously, the zone is typically on the order of 1,000ft (300m). In this region it is beneficial to be equipped with a gun as well as AAM. Beam quarters are clearly more challenging for fuzing and warhead effect given an annular pattern (Low Fuzing Probability zones), with less likelihood of causing lethal damage if the warhead is detonated at all. This zone aft of the non-maneuvering target represents inadequate closing speed for, again, low-probability

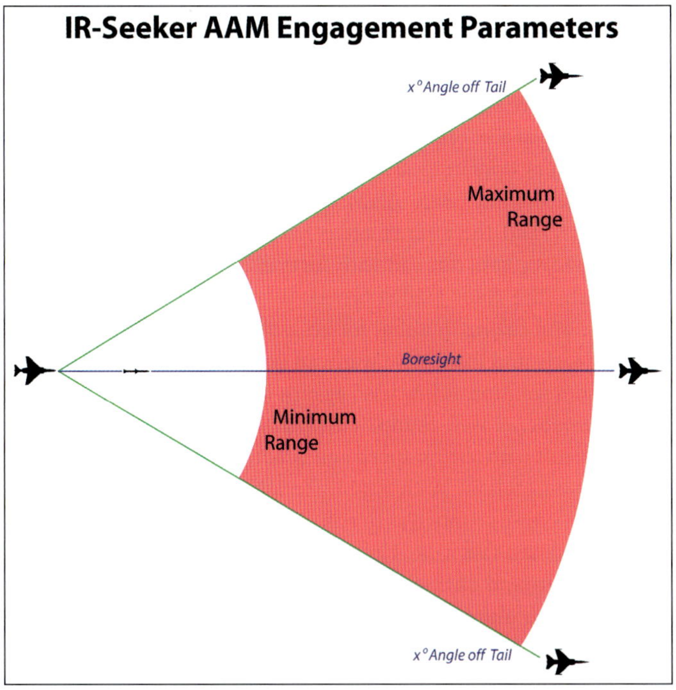

Figure 59: Conceptual top view of a non-maneuver target engagement envelope for an early heat-seeker missile. Angle-off-tail is a limitation of the missile seeker head and design of the target aircraft that might provide some masking of the exhaust at certain angles off the tail. (Drawing by author)

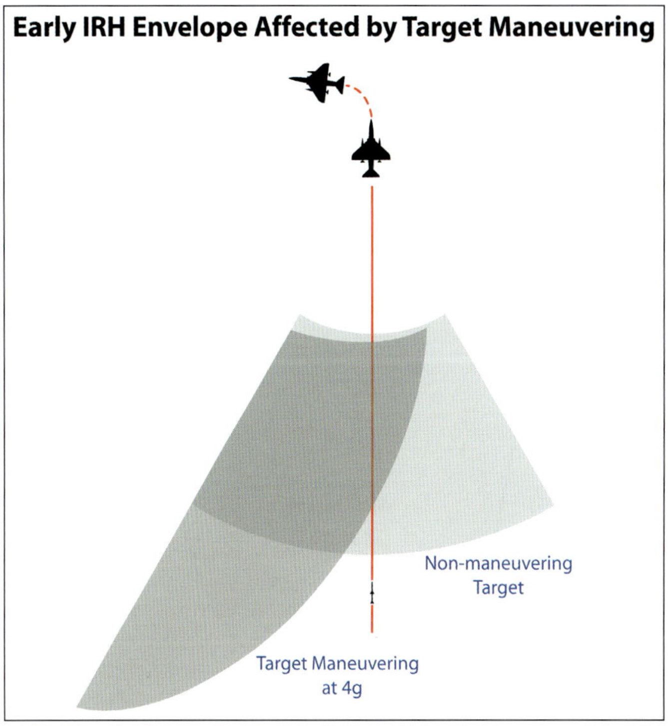

Figure 60: This is a basic look at how the top view engagement envelope for an early IRH missile is skewed by a turn. Because the target piloted is cutting the relative speed and distance by turning, shortening the AAM's flightpath, AAM range is effectively increased to that side. The span of angles available for the seeker to lock onto the exhaust is, however, greatly reduced. (Drawing by author)

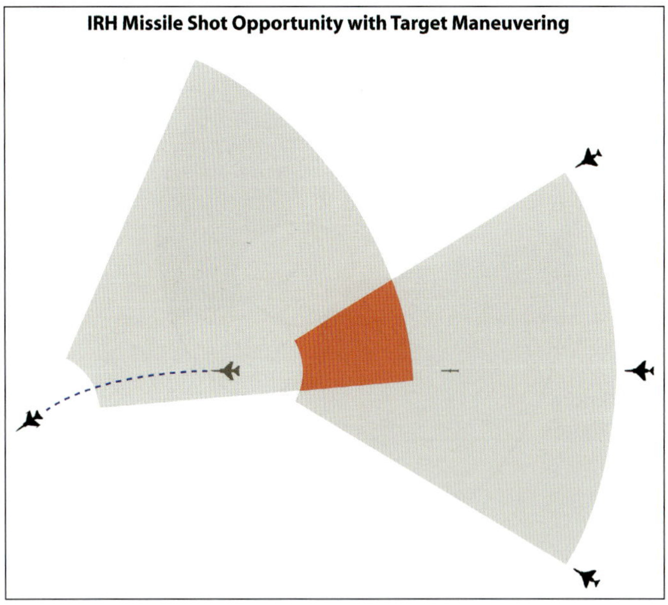

Figure 61: This is another look at the effective engagement cone for the early generation IRH for a non-maneuvering and maneuvering target, graphically depicting how the shot opportunity shrinks dramatically with even a modest turn by the target aircraft. (Drawing by author)

fuzing. The Look-up Required zones, superimposed on the Low Fuzing Probability, reflect the performance of a ARH missile and its performance in the presence of ground clutter. The target is unlikely to be detected in these beam aspects except against the sky.

For the maneuvering target, in a steady turn throughout the engagement, the shooter on the inside of the turn would be seeing the top of the enemy aircraft but the bottom when on the outside of the turn. As shown in Figure 63, this offers both advantages and disadvantages given that the target is either turning into and across the shooter flightpath (closing on a missile in flight) or turning away.

Engagements close to the ground add more problems with missile dynamics potentially leading to ground impact and difficulties with radar and IR reflections off the surface, The lower hemisphere for maneuver is largely eliminated.

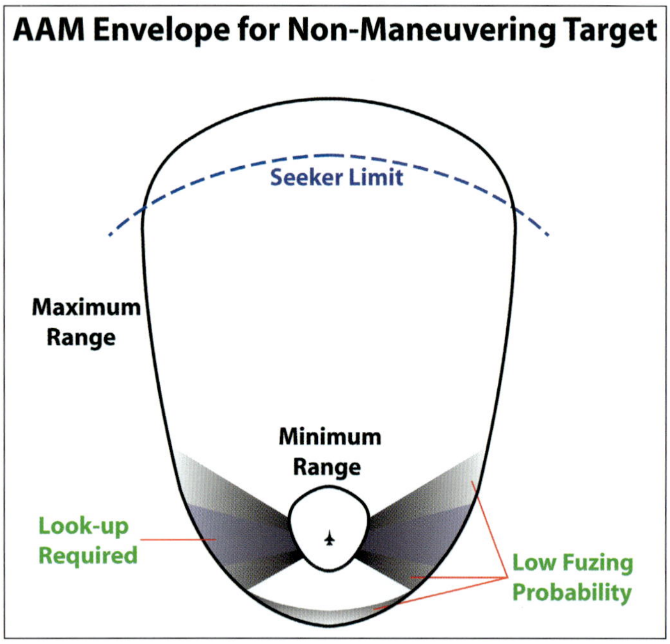

Figure 62: Depicted is the conceptual top view AAM engagement envelope for a non-maneuvering target. (Drawing by author)

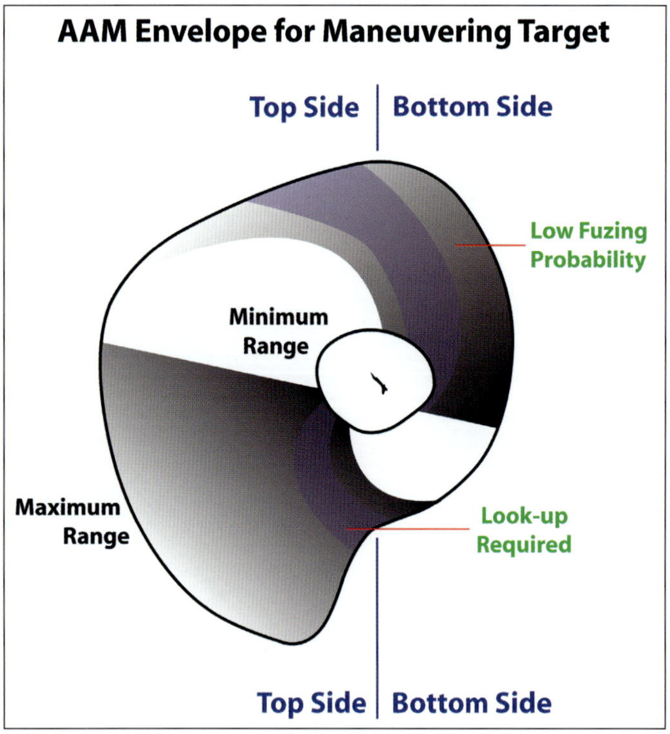

Figure 63: Depicted is the conceptual top view AAM engagement envelope for a maneuvering target (constant level turn). (Drawing by author)

These diagrams would differ for different aircraft, missiles selected, flight conditions, and maneuvering (especially aspect). Some elements may be displayed to the pilot by the fire control computer. All must be considered by the firing pilot, depending upon circumstances, to fire with high probability of downing the adversary, and the defending pilot seeking to survive the engagement. Even with advanced, high-speed onboard computers and sensors, a certain 'shoot cue' to the pilot for every circumstance has never been achieved, and likely never will. To reduce the complexity, some 'rules of thumb' are generally memorized. Much still relies on pilot (fallible) judgment of range and angles matched to knowledge of weapon capabilities. Purely analytical computer modeling can be performed for the warplanes and AAM likely to meet, especially ahead of a known potential conflict, to determine these rules of thumb. Better yet are learning how to gain advantages through dissimilar air combat training (DACT) and pilot-in-the-loop simulations in which aircrew 'fly' against opponents in linked simulators. All of this is dependent upon the fidelity of the training and the reliability of the data on which the opposition systems are modeled (based on intelligence collection and extrapolation). In actual combat it all requires a cool head, situational awareness, system knowledge (especially limitations), and adaptability. Aircrew proficient in all these skills is the difference between a fighter pilot and a heavy equipment operator – or a target, from the opposition's perspective.

All of this emphasizes why so many AAM shots 'miss', apart from system malfunctions.

Difficult Targets

Modern AAM have been designed to engage jet aircraft like bombers and fighters with good IR and radar signatures. Aircraft with reciprocating engines are much more challenging given their low heat exhaust and propellers reflecting confused radar returns. Even turboprop aircraft would be difficult targets. Cruise missiles flying at exceptionally low altitude and UAVs with small size and engines are equally challenging. The fifth generation of AAM generally had seekers sensitive enough to place all such targets at risk, but still with less certainty of a kill than the typical jet combat aircraft.

Helicopters operating at low altitude are especially difficult to engage successfully with AAM. The turboshaft engines operate even cooler, and the aircraft often have shielded exhausts mixing in surrounding cool air, and with further cooled by the main rotor downwash. The rotor blades present a confused picture for radar guidance and fuzing, making for an erratic missile flightpath with possibly early and ineffective warhead detonation. Just trying to get a good shot solution at low level against a close ground background is problematic. A well-trained chopper pilot can make this all the more difficult with abrupt speed and heading changes impossible for an airplane to follow. With a gun, the attacking pilot must depress the aircraft's nose so much as to risk collision with terrain. An AAM, at least, reduces that risk. However, missiles have been known to impact the ground during their maneuvering as they rarely (if ever) contain a radar altimeter and terrain avoidance computations in the guidance solution.

Vulnerability

The targeted airframe under stress from flight and maneuvering airloads can readily fracture with even small cuts. This can produce a cascade of structural failures and breakup of the aircraft or loss of

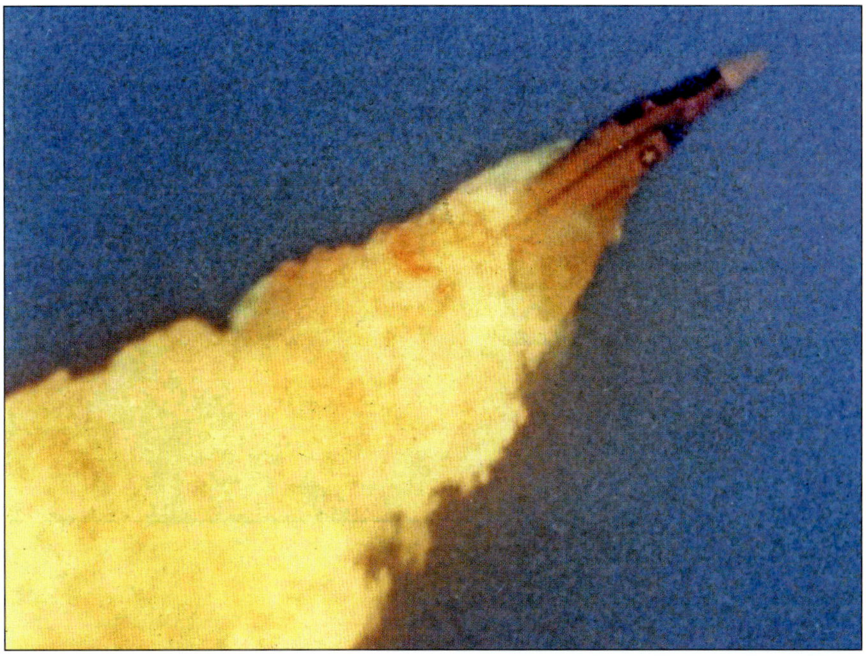

A QF-4 drone is nearly fully engulfed in flames from burning fuel, hydraulic fluid, and other combustible elements after an AIM-54C missile hit in 1983. Combat aircraft are designed to reduce the potential for fatal damage from an AAM hit and subsequent fire and explosions, but some damage is too severe. (National Naval Aviation Museum)

control. Penetration of fuel and hydraulic system components can induce a fire and explosion. Damage can also cause systems to malfunction in addition to disabling or killing the aircrew – the 'softest' element of the aircraft. All are effective in destroying the aircraft, forcing the aircrew to eject, or causing them to withdraw for an emergency recovery.

Design guidance to make aircraft more survivable to an AAM hit is evidently no different than general 'hardening' to survive gunfire, AAA and SAM hits. Some systems and structure are more vital than others, and damage-tolerance design practices help mitigate vulnerability. The principles are system redundancy, avoiding critical items routed or mounted in close proximity to each other, fire suppression systems, and armor (heavy) where critical. Aircrew seats of combat aircraft commonly have some armor to reduce the potential for fragments reaching the individual. However, irrecoverable damage to the aircraft, especially if it breaks up or explodes, has high potential to fatally injure aircrew unless they have egressed via an ejection seat.

The potential exists that the vulnerability of the aircraft is increased by carriage of AAM should the missile be hit by hostile fire (shell or warhead fragment) such that the motor or warhead explodes. The author has found no evidence of such an event except for an AAM catching fire after the aircraft was hit by AAA.[7] Particularly solid propellants need a very intense flame to ignite, and warhead explosives are not efficiently ignited in this manner. However, there may be many other inflammable components of an AAM, depending on design. Certainly, situations such as a store of rounds in an aircraft carrier magazine subjected to persistent fire

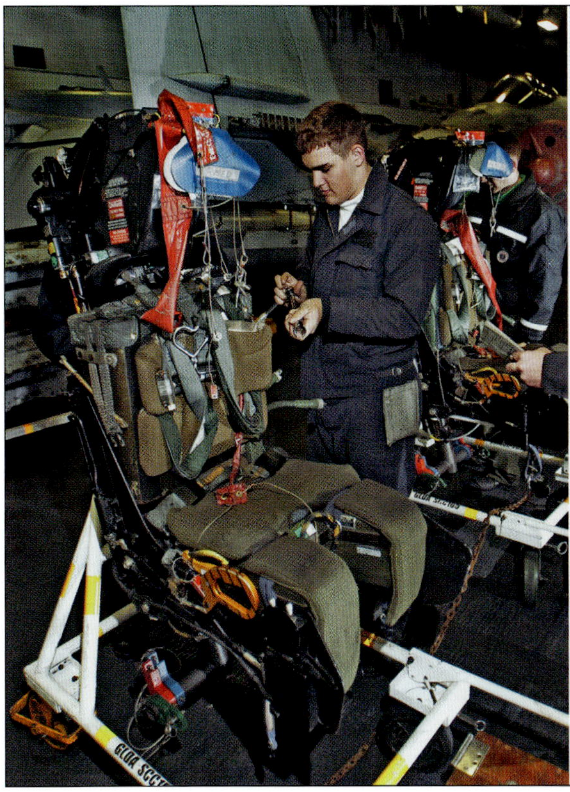

At left a US Navy mechanic services an ejection seat from an EA-6B aircraft. The yellow handles beside the thigh cushions initiate the ejection with a rocket motor propelling the seat up a rail and a parachute in the container behind the headrest deployed at the right moment to slow the descent. The aviator releases from the seat at a safe altitude and deploys a personal parachute. At right a pilot ejects from an F-16C during an airshow in 2003, realizing he was too low to recover from a Split-S maneuver. Shown is the intense flame from the rocket motor and the canopy separated at initiation of the ejection. (US Navy and US Air Force by SSgt Bennie J. Davis III)

The pilot of this F-20 prototype is firing an AIM-7F from an inboard station. The large and smoky exhaust plume is going to pass close to the side inlet for the engine, illustrating one purpose of the test. It is necessary to ensure that if the engine ingests the plume, it does not stall or, if it does, it recovers quickly. (Author's collection)

can 'cook off', but on an aircraft in flight this is much less likely; the jet would have been abandoned by that point. Specifications work to ensure the design of explosive components are unlikely to detonate under less than extreme circumstances. By 2010 this was referred to as 'insensitive munition' standards. Tests are conducted to demonstrate compliance.

The launch aircraft jet engine(s) may ingest the rocket plume of the AAM and suffer a compressor stall (internal airflow disruption) or flameout. This concern prompted Soviet authorities to prohibit R-3S carriage on MiG-23 fuselage rails after flameouts were experienced. The engines on the latest fifth generation fighters are designed to self-recover from such an event while in the past the pilot had to work to achieve this or initiate an air-start sequence amidst an air battle. If the restart is unsuccessful the aircraft would need to be abandoned if a single-engine aircraft. Location of the missile launcher, ignition only after ejector launch and so clear of the aircraft, and tactics (training) can mitigate this potential.

For close engagements it is possible for the launch aircraft to inadvertently fly through debris from the missile detonation and the target which has been hit. Aircraft have been damaged and even brought down by such collisions. At least F-15 pilots were taught to roll away after firing a Sparrow AAM, in what is called an f-pole maneuver, to create maximum separation between the launch aircraft and the target at intercept.[8] Such maneuvers can also ensure vision unobstructed by the smoke from the AAM motor. Also, during AAM fly-out, the launch aircraft for a SARH missile is vulnerable to intercept. It is often wise to break off after the shot because another adversary may have closed while the shooter was preoccupied. Some systems do not permit maneuvers that would break radar lock on the target, and so the shot may need to be abandoned if the launching aircraft is at imminent risk.

Failures

Missiles are complex and have many failure modes that can cause them to malfunction and so not perform as required. The choice of propulsion type, sensor type, fuzing, and many other aspects can increase complexity and so potential failure modes with reduced reliability. Especially early electronics were fragile and even later designs subject to component failure under prolonged vibrations induced during captive carry flight. Malfunctioning electronics can affect the seeker head, guidance and control, and fuze segments. The rocket motor may also fail to ignite or, if the propellant has degraded in storage, perform poorly or explode on ignition. For liquid-propellant missiles the many pumps and valves can also fail or operate non-optimally owing to degradation, aside from fuel leaks that can lead to fire and explosion. Ground crew can install the missile and make the connections to the aircraft improperly (or not at all in the case of the connections). There is typically an individual monitoring this operation to check for such errors and aircrew should look for these on their preflight 'walk-around.' Systems on the aircraft side can also fail or, in the case of services delivered from the aircraft such as seeker head coolant, the supply may leak or deplete normally. Even on the missile side things like batteries drawn down can negate operation.

Even with the most reliable components, say 95 percent confidence of proper functioning, the cumulative uncertainty for all components taken together can drive down overall probability of destroying the target. This is risen if one failure promotes another in a cascading effect – a dependent versus independent failure. If, for example, there was 95 percent reliability for each major action of an AAM's success – the aircraft sending a suitable firing command, the missile launching, then guiding, closing on a non-cooperative target, fuzing and then detonating – the cumulative probability of a kill is just 73 percent if these are independent failures modes. History, at least up to fifth generation systems, would suggest that anticipating more than 50 percent reliability is too optimistic. In general, IRH missiles have been approximately twice as reliable as SARH and ARH systems. This low reliability is what drove salvo shots of more than one missile during the early decades of AAM use in combat and dictated by tactics. For example, firing two AAM, each with an individual 73 percent probability of kill, raises the overall probability to 93 percent, all else being equal.

Early systems had a basic check of missile readiness, indicating ready or fail to the aircrew. By the twenty-first century, with greater onboard digital processing, it became possible for the missile to perform self-initiated BIT to check readiness of multiple systems and individual elements prior to indicating readiness for launch. The cause of a failure can be stored in memory for download after landing by armament personnel for diagnosis and to guide remedy. Tracking such failures can identify the most problematic systems for possible improvement in the next block of upgrades, raising overall fleet reliability.

Countermeasures

Countering an AAM begins by becoming aware of the attack, either from direct observation of the launch and/or fly-out, a radio call from someone else observing it, or by indication from an onboard sensing system. Missile Warning Systems (MWS, or Missile Approach Warning Systems, MAWS), to detect an incoming missile, became available around the year 2000. Detection is performed optically, by observing the intense rocket plume. Sensors on the aircraft detect the emissions in multiple portions of the spectrum (principally IR and UV) and the system determines the source via comparison of this signature with those saved in memory. The bearing and elevation of the threat can be determined by comparing the output of multiple sensors installed around the aircraft for full-spherical coverage. (The lower hemisphere is typically emphasized for transport aircraft defense against SAMs.) The aircrew is alerted, usually visually and aurally, and the data passed to any onboard defensive systems (see later).

To become aware that launch of specifically a radar-guided AAM is imminent or has been executed, a suitably tuned RWR (Figure 64) on an aircraft will alert the aircrew that they are being 'painted' and locked-up by threat radar. This usually requires the most likely adversaries to be identified ahead of time so that the RWR can be tuned to those frequencies and pulse repetitions. Such information is typically derived from intelligence collection (RF intercepts and analyses). Initially, RWRs had only a few such channels, but this increased apace with computer technology such that a whole database of such characteristics can be stored and more in-unit analysis performed.

Knowing the characteristics of foreign AAM, especially if upgraded, lies in the field of intelligence collection or espionage. If an enemy AAM was supplied by a friendly nation, obtaining the desired data is a matter for diplomats and military attachés. Should a foreign product become available for close examination and test, most militaries have organizations assigned to such tasks. Updating and disseminating this classified information ahead of an anticipated conflict is a vital role.

The most basic means for a pilot to avoid being engaged by an AAM is to avoid detection or dissuading the opposition pilot from firing in the first place via passive measures. The former is done by avoiding being seen visually or via radar, remaining out of range (having knowledge of opposition AAM assets), presenting an unfavorable aspect, drawing the shooter into unfavorable maneuvering flight conditions compared with the target aircraft, and confounding radar and/or missile lock. Just drawing the fight to very low altitude can present greater difficulties for the shooter though also limiting options of the defender and placing the aircraft within range of ground-based air defenses. Many of these choices require knowledge of the enemy systems and situational awareness of the battlespace. Experience, training and a cool head counts for much in this game. Failing these, countermeasure systems can

An example RWR is this Russian SPO-15. A light in the three arcs of indicators around the aircraft planform symbol would indicate the bearing of a threat, each arc presumably some characteristic of that threat. The bottom row of lights is programmed to indicate the nature of the threat. (via Ali Tobchi)

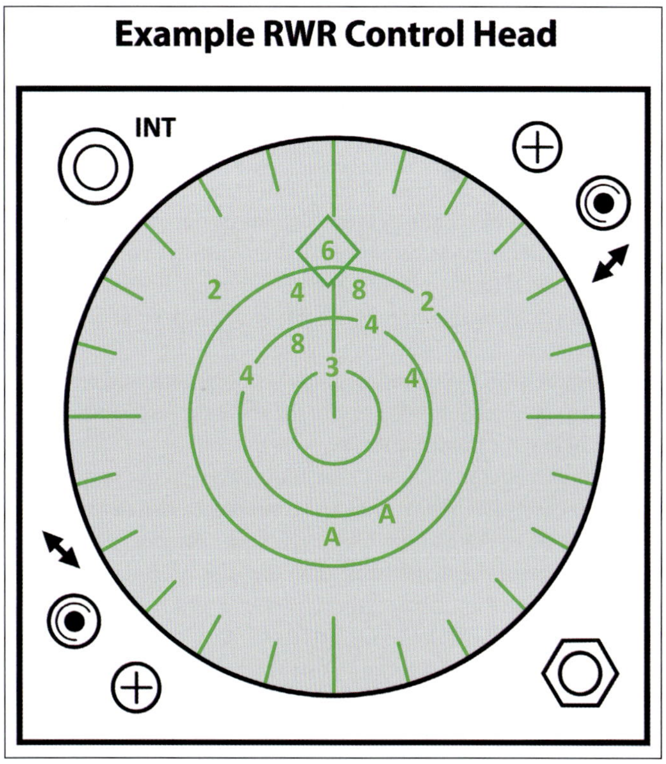

Figure 64: Another example RWR control head and display is illustrated. Controls include display intensity (a green raster scan shown) adjustment of bearing via two knobs, and the display identifying emitters via alphanumeric codes with one boxed to indicate priority. (Drawing by author)

be employed. Uncertainty remains in the type of missile fired, its reactions to the CM and their effectiveness.

The easiest and most practical means of countering an AAM is the 'break', which is an abrupt roll and pull into a high-g turn (a rolling pull-up or pull-down). If performed at the optimal moment, this seeks to break the seeker lock or create maneuver requirements the AAM cannot meet owing to depleting energy. The early AAM could not follow a target maneuvering beyond a few g's or outside the narrow FOV of the seeker. Consequently, a roll and pull away from the threat could be executed in a second and had reasonable potential for disrupting the lock of the missile. Even for later more capable generation missiles, the break remained a good (and possibly only) countermeasure, coupled with systems-intensive means discussed in the next section. The maneuver increases crossing speed (LOS rate of change) and so challenges the tracking ability of the AAM. Breaking in the direction the missile is approaching – that is across the path of the missile – most quickly establishes that LOS rate of change. For a rear quarter this is turning into the missile and forward quarter turning away. After a 90deg heading change, assuming the missile is now in a beam approach, a reversal may be advisable to cause it to make a large correction to reestablish a lead-collision intercept. All this requires awareness of the direction of missile approach. The break also alters the aspect angle to what may be less favorable to a successful engagement – high crossing speed, more challenging fuzing scenario, non-optimal warhead fragments dispersion. It may be easier to defeat a missile in a head-on engagement as it may be visible to the pilot whereas seeing an AAM closing from the rear, and judging proximity, is more difficult. A break also allows the pilot to look back for the source of the missile closing from an unseen direction and perhaps plan further evasion or engaging the opponent to advantage. Again, timing it vital. The break can be ineffective if performed too soon, the missile suitably adjusting its flightpath, or too late with endgame already determined. It is often an agonizing wait by the pilot with a missile inbound to make the break at the best split-second moment. If the missile is detected far away, placing it astern while increasing speed can also serve to outrun the round until it loses energy and falls away.

When aircrew observe an AAM closing on a friendly, they are trained to give the targeted aircraft's call sign when transmitting a warning. A call like 'Elmo 2, break!' avoids having every aircraft on the mission frequency execute a break simultaneously, disturbing the formation plus risking midair collisions. Giving a direction to break might also be advisable if judged – 'Elmo 2, break right!' However, the excitement of combat often overcomes training. If the inbound AAM can be seen by its smoke and it is still guiding and closing, it is time to try something else. Some aircrew have just panicked and ejected before the missile arrived.

A targeted aircraft aircrew, or its automated systems, may employ countermeasures to defeat the AAM intercept. The development, fielding and upgrading of CM occurs in parallel with AAM. Countermeasures such as flares, chaff and decoys frequently must be 'tuned' to the anticipated threat systems and employed at the right instant to be effective. This last has come to demand automatic sensing of the threat (detecting radar 'painting' or missile fly-out) and automatic deployment of the CM. A slow-moving chaff cloud is relatively easily discounted by a Doppler radar tracking system.

Countermeasures seek to defeat the AAM by evasion of the missile or confounding its tracking of the target. Several means have been developed and evolved over the decades. The missiles themselves have, likewise, been improved to counter these methods via CCM in an endless cycle. The goal is to maintain an advantageous jamming-to-signal (J/S) ratio to remain effective by defeating the missile. Hence, combat may occur in a period, and between opponents, when one or the other is CM/CCM deficient and so either the shooter or defender holds an advantage. Other combat means must be brought to bear to compensate. As AAM became software and digital processing intensive, the ECCM involved a 'library' of data to identify threat systems and determine the optimal CCM to reduce J/S significantly. The library, embedded within the software, is typically classified and not made readily accessible to other nations who may purchase the AAM. This is one reason other operators seek access to the computer code, to be able to update and optimize the CCM to their current needs. Denial of this sharing has been cause for international disharmony and cancelled contracts. It is one rationale for nations to develop their own AAM.

Radar Homing

Evading the missile with a break is possible, as describe previously. Turning to present the lowest radar reflection may be practical if the bearing of the inbound AAM is known, and a continuous turn to create varying glints may confound the seeker. Repeated maneuvers, which the AAM attempt to follow, will deplete its energy, cutting its range and endgame potential. Performing a downward break (with sufficient altitude) by rolling far enough to pull down (requiring a few more tenths of a second) helps to maintain airspeed (with the advantage of the pull of gravity) and presents a look-down perspective for the guiding radar with the potential for ground clutter to confuse the tracking. Furthermore, in flying a lead-collision course, the AAM will be guiding to a point that may be below ground level. When the pilot pulls up just above the terrain the AAM may be unable to correct in time and 'smites the earth.' For a radar reliant on Doppler Effect to detect the target (pulse-Doppler), just maintaining co-equal airspeed (with an aircraft guiding a SARH missile) to reduce potential for a sustained lock can confound the radar tracking. This can be done by flying perpendicular to the attacking aircraft or ARH missile to reduce closure rate in the 'Doppler Notch' (the perceived speed the same as passing terrain). This is generally only effective when flying against a terrain background. However, 'just sitting there' is a counter-intuitive tactic for a pilot under fire.

A passive RF countermeasure is to reduce the radar reflectivity (RCS) of the aircraft via design and/or external coatings. While some small measures can be taken as modifications to existing aircraft, the physics demand greatly lowering RCS to significantly reduce radar detection range. Such 'stealth' characteristics are typically achieved by radical external design configurations, construction material choices, radar absorbent material (RAM) coatings, and eliminating all external protuberances most especially carrying all stores internally. At best this can reduce RCS in certain aspects viewing the aircraft – generally to the front – but rarely all aspects. Reducing detection by ground-based radar has been the principal objective, and this does not necessarily translate into reduced vulnerability to air intercept radar. Application of RAM to existing aircraft has generally sought to reduce glints and reflections from acute corners. Such areas as the engine intake, engine exhaust, and cockpit cavity are generally impractical to treat in this manner. Just the radar dish is a prefect radar reflector at certain frequencies. Since RAM will degrade over time through wear-and-tear, especially if it is applied as a paint across large areas, it is important that it be inspected sortie-to-sortie and refreshed as needed.

Lowering the RCS of an air-to-air missile is generally considered only in reducing the RCS of the launch aircraft for external carriage of the rounds. This can be done, as with the aircraft, by outer moldline shaping, selection of fabrication materials, and application of RAM. The author is unaware of any fielded AAM with these characteristics. The compromises in missile performance have been considered a poor tradeoff. Only after 2010, when conceiving sixth generation VLRAAM of large size, has reducing RCS to preclude detection by anti-missile systems been considered reasonable.

Active CM is jamming or spoofing the guiding radar on the launch aircraft or the AAM. Jamming requires RF signal analysis and dedicated transmitters plus antennas backed by high-power electrical sources onboard the aircraft or ground station. Knowledge of the threat systems, such as transmit frequencies and PRF, will inform the optimal ECM to prepare, such as emission modulation. The jamming power must be high enough, compared with the source radar, to create a detrimental J/S to cause loss of lock. Presenting a signal of greater strength than what would normally be returned, but altered to divert the tracking, can misdirect (spoof) guidance. Jammers were initially small enough, with modest power and volume demands, to be mounted within a fighter. These rapidly grew to external pods, some with internal power generation rather than supplied by the aircraft. By the fifth generation of fighter, especially seeking reduced RCS, the jammers were back inside the aircraft. For radar-guided AAA and SAMs, jamming over a large area has typically required specialized mission support aircraft equipped with multiple jammers and able to generate the high electrical power required. These have contributed nothing to air-to-air combat, but it is not inconceivable they could be used to defeat VLRAAM inbound to high-value targets.

A CCM to jamming is the SARH or ARH home-on-jamming capability. Defeating a HOJ engagement (if detected) requires the jamming to be cut off or sufficiently altered, though this could then permit the AAM to resume its normal homing.

Another tactic is to decoy the radar-guided missile away from the aircraft with a larger reflective glint. By the late 1960s this had come to consist of dispensers ejecting bundles of chaff that open and spread out in the airstream. The chaff is typically long and narrow strips of metal foil, or metallic coatings on some thin paper or plastic material. The resulting cloud of chaff reflects the radar energy as a larger target to which the AAM will hopefully guide. It is important that the length of chaff is closely matched to the wavelength of the guiding radar and so, depending on anticipated threats, the appropriate chaff 'expendables' are loaded in the dispenser. The aircrew can 'punch off' the chaff manually or rely on a computer to select the appropriate expendable and dispense it at the right instant such that it does not disperse too soon. The missile may have processing that helps to discriminate between chaff and the real target based on differences in the reflection and relative motion. Especially if guided by Doppler Effect, the relative motion of the target and the chaff cloud (that can also be blown by winds aloft) will determine the effectiveness of this decoy method. However, if the missile's Doppler Gate is set too narrow in an effort to avoid ECM decoying, it is more likely to lose lock for more typical radar tracking reasons.

A flying decoy offers a more realistic alternative target with large radar return or emitting jamming signals. Even if the AAM seeker alternates locking-up the target and the decoy the varying guidance commands, with resulting jerky flightpath, can drain maneuver energy and range. If the AAM homes on the false signal, then it only destroys the inexpensive decoy instead of the combat aircraft. The decoy can be free-flying darts, powered air-launched UAVs, or small aero shapes towed on a line from the aircraft with capability to reel it back in for reuse. A trailing decoy is set far enough back to ensure against damage to the aircraft by warhead detonation. Trailing or

This Israeli F-15B is equipped with a locally manufactured Ophir photo-reconnaissance pod on centerline. It is heavily armed for the mission with what are likely three AIM-7Ms and four Python 4 missiles. On the forward port AAM mount of the conformal fuel tank is an Elta EL/L-8212/22 electronic warfare pod. The ideal is to execute the mission without firing any weapons, and EW resources may be vital to that end. If a fight becomes unavoidable, those same resources could help to ensure a successful outcome. (Ofer Zidon collection)

A typical countermeasures dispenser can look like these four in the side sponson of a C-130 aircraft, some expendable cavities unfilled. The expendable can be chaff or flares. Round cross-section expendables versus these square types are also common. Generally, the color and markings denote the unique characteristics of each expendable. (USAF)

launching a decoy with higher radar reflectivity like a simple radar reflector may also be effective.[9] Small decoys may be contained within the aircraft, such as a bomber, to be dropped or deployed on the trailing line. Fighters would normally carry the decoy as an external store or trailed from an external pod. Stealthy fighters, for which external stores are anathema, have been designed to deploy the decoy from within the airframe. Generally, just a few decoys are carried and when all are destroyed other CM must be brought to bear. However, the typical combat mission would experience just one and maybe two intercept attempts with one AAM fired during each.

IR-Homing

The easiest and most practical passive means of defeating an IR-guided AAM is the break, as already described. For early tail aspect limited AAM this has the added advantage of greatly reducing the engagement envelope by masking the hottest parts of the engine exhaust. This can be amplified by a power reduction (especially coming out of afterburner) and maneuvering such as to use elements of the aircraft to shield the exhaust. Even all-aspect AAM could be challenged by having to guide on less intense IR emissions. Pulling across the sun or sun-reflective surface like the ground also has the potential to induce a loss of lock. Flying into a cloud can obscure and scatter the IR energy with the potential for a loss of track. However, with IRH systems generally being short-range weapons, with flight times around 10sec, the time for identifying a missile fly-out, deciding a course action, implementing this, then aircraft reaction time can make the margin for such measures all too brief. The advent of IIR seekers with staring (non-spinning) FPA, on a suitably fast gimbal mount and possibly with flight path predictive software, reduces the effectiveness of maneuvering to break lock.

Another passive countermeasure is by design of the aircraft to reduce the IR signature of the exhaust. The targeted pilot can pull power to lower the exhaust intensity, but this has no immediate effect on lowering the temperature of the exhaust structure that cools slowly. Aircraft design can lower the inherent IR signature of the exhaust via shielding and by entraining cooler surrounding air to help cool the exhaust structure. This has generally been exceedingly challenging owing to the effect of intense heat, pressure and acoustic load of the exhaust on surrounding materials plus possible thrust degradation.

The active, more electronic-intensive IRCM are intended to defeat the AAM by having it track a decoy, if even briefly, such that the real target leaves the seeker IFOV and is not reacquired on a subsequent scan of the FOR. The IRCM can come in several forms. These all seek to create an IR source more intense than jet engine exhaust, even in afterburner, to confound missile tracking.

Among the earliest forms of IRCM, and persisting, are flares released from the aircraft. Initially these were ejected from an externally mounted dispenser described in the previous section. Stealthy aircraft have come to use internal dispensers exposed briefly by rapidly operating doors. Flares became available during the 1960s and have evolved. These were initially pyrotechnic devices, combusting without the need of the oxygen in the air. Magnesium was a common material for this purpose, heated to incandescence. In the latest decades these 'flares' became coated metal wafers for pyrophoric release of energy – smoldering spontaneously when exposed to the air's oxygen and releasing IR energy with little visible light. The decoy falls away from the aircraft and so decoys the missile seeker. All are designed to emit in a specific bandwidth and with rapid intensity rise time. As a 'rule of thumb', the flare intensity needs to be twice that of the target to be effective in decoying the AAM seeker. As missile CCM improved it became important that the flare's IR emissions closely match the IR bandwidth of the engine to avoid it being discounted by the discrimination techniques. Several selections of IR flares can be loaded based on the anticipated threats. For all-aspect and multi-color IRH systems the flares have had to cover an even broader bandwidth. Some of the advanced CCM discrimination includes size of the IR source, rise time, and trajectory not associated with the aircraft as it is tracked. Thus, it is also important that the flares be ejected forcefully such that they

This ALQ-184 jamming pod (seen from the rear) on a US Air Force F-16 contains integral towed decoys. This is indicated by the flared dispenser below the aft end antenna with four covers. An example towed decoy is the BAE Systems AN/ALQ-55, seen in the inset image, on a fiber optic line that likely carries electronic signals. So, this is an active decoy rather than passive. Surfaces have popped out to stabilize the decoy that is less than half a meter long. (Raytheon and US Navy)

An aircraft exhaust can be designed to reduce the IR signature by shielding, reducing the angles from which the exhaust can be viewed, and mixing with ambient air for cooling. These Israeli AH-1 Cobra attack helicopters model two variants of engine exhaust IR signature reduction. At top is the 'sugar scoop' exhaust diverter to turn the hot flow up and so reduce the intensity of the signature visible to a man-portable air-defense system (MANPADS) sensor below. At bottom is an extended exhaust that allows the exhaust to cool before reaching the exit. The pipe is also covered with a woven metal mesh that reduces the intensity of the heat reaching the outer surface. (Author's collection)

Exhaust shielding was done with the F-117 shown by squeezing the exhaust to a slit, using vanes to shield the view of the turbine plus also helping to mix the exhaust with surrounding cooler air, and providing additional shielding via the extended lower lip. Some of these features were also to reduce radar signature. All this compromised aircraft performance and required considerable maintenance attention but was accepted for the exotic aircraft with very specific mission profile. The extraordinary F-117 external moldline reflects early RCS reduction principles and also had RAM applied. (USAF)

move rapidly and laterally away from the aircraft so that, with the missile closing rapidly, it is suitably decoyed. The very narrow IFOV can also mean that flares are missed during a scan, and so the routine has become to eject several flares sequentially, sometimes at a specific rate to match the scan rate of the anticipated opposition systems.

Combining flare ejection from one side of the aircraft with a break in the opposite direction can improve the chances of successfully decoying the AAM. The pilot might also retard the throttle to reduce engine IR emissions during flare dispensing and the break. Depending on the fuzing means, the missile may never approach the real target close enough to detonate. Aircrew may also deploy flares in anticipation of a missile launch so that the seeker selects the flare initially. That is more practical when in a SAM-rich environment, the pilot popping flares climbing out of a ground attack pass. In general, flares are more effective in defeating SAMs than employing them in air combat. When the AAM may be just seconds from launch to impact, choosing the best moment to eject flares (possibly a fraction of a second) is very challenging. The system detecting the inbound missile via the motor plume's electromagnetic signature can dispense multiple flares of the correct type at calculated optimal intervals. This will continually cause seeker deviation and reduce the chances of reacquisition of the intended target until the AAM has exhausted its energy and maneuvering potential. Manually 'popping' the expendables has lower chance of success – although aircrew may do it just for a sense of 'doing anything.'

The advent of MWS was driven by the growing threat of man-portable air-defense systems (MANPADS) or a 'shoulder-fired' IR-guided missile that lies outside the scope of this text. The detection and automatic response require that the missile launch be detected, as described previously. The challenge is that the rocket motor will very likely burn for only a small portion of the missile's flight after which tracking it becomes much more difficult. Hence, the MWS and any defensive systems must perform with very high reliability and detect the launch within the first second or two of flight and then respond with quick-acting response such as dispensing flares and chaff. This response can be programmed with cockpit control head settings. The MWS computer can select the optimal flare based on programmed logic, and dispense the expendable judged most suitable. Such systems are more typically installed in attack helicopter and transport-class aircraft countering SAMs than fighters.

Infrared 'flashers' use lamps to emit bursts of IR energy, usually in the aft quarter of the aircraft, to confound the IRH seeker. This does not require a MWS because the system is simply turned on when entering a combat zone. A very high-energy burst can temporarily blind the seeker, but a more practical approach is to confuse the device. The flashes can be modulated to match the rate of the missile's reticle scan rate, such a 120Hz, or sweeping a frequency band in a number of predetermined patterns. This is especially effective against IRH systems with spin scan, but even imaging systems that sample pixels in a repeating fashion can be vulnerable. For others, even occasional jamming during the fly-out can cause energy-depleting maneuvering that will cut range and possibly induce a miss. The earliest form of this employed a lamp at the aft end of an externally mounted pod that also contained the electronics and a power source such as a nose-mounted propeller-driven generator. These appeared during the 1970s. They were quickly rendered ineffective by advances in seeker head design, especially frequency modulation that made the flasher the bright target rather than a countermeasure. A later system appearing on American combat helicopters, the AN/ALQ-144 'Hot Brick' jammer, has a block of silicon carbide heated to incandescence as an IR source surrounded by a spinning cylinder of lenses sending the light out in multiple directions simultaneously. Later systems use

An F/A-18 pilot pops flares while executing a break turn, the airplane and the countermeasures quickly separating in opposite directions. The brightness suggests flare emissions in the visible wavelengths as well as IR. Flares can be dispensed automatically by an onboard system or 'popped' manually by the pilot. (Public Domain)

lasers emitting in the IR band and sending the light out via a rotating mirror directly at the AAM. Even though emitted by the target, the flashes cause the missile to perform sharp flightpath corrections that can, ultimately, place the true target outside the very small IFOV of the seeker. This breaks lock that may not be reacquired if the flashing is maintained, repeatedly confusing the seeker. The more modern IIR seeker, with extensive digital processing, is better able to discriminate the light of such IRCM and reject trajectory changes inconsistent with tracking of the real target. These are better countered with towed IR decoys that move as the aircraft moves but emit more intensely, though their emissions must match the aircraft as closely as practical to be effective.

As flares and lamps became increasingly ineffective against newer missiles, especially employing IIR, new IRCM means were sought. Solid-state lasers were adopted, aimed by computer towards the missile sensed by the MWS. A high-intensity laser light, usually equating to power available, is necessary to ensure a suitable J/S ratio. This can overwhelm the seeker, 'dazzling' or blinding it, and if of great enough intensity, can 'fry' the sensitive electronics. The laser light is emitted from a small turret, aimed by the MWS, projecting from the aircraft. Initially a tracking laser is aimed at the threat and sensed reflected laser light off the AAM's irdome is the indication of acquiring the target and it remains locked via closed-loop tracking. Seeker modulation can also be detected via this laser for aligning

This Israeli Blackhawk is fitted with a 'Hot Brick' IR jammer atop the engine bay, aft of the rotor, as well as two large countermeasures dispensers of expendables farther aft on the fuselage side. Some of the various small black antennas installed around the helicopter may be for a MWS. Combat helicopters have traditionally been more concerned with MANPADS threat, but the ability of modern IR-guided AAM to engage low-flying targets like rotorcraft has increased that risk. (Israel Air and Space Force)

The long gray pod beneath the starboard wing of the flight test A-37 is a Loral AN/ALQ-123 infrared jammer generating high frequency IR flashes out the aft face. At the front is the propeller of a ram air turbine to generate electrical power. Correctly modulated, the IR flashing can confuse an IR-guided missile and cause it to lose lock. (Author's collection)

The circular assembly beneath the aft fuselage of this US Air Force C-17A is an AN/AAQ-24(V) Large Aircraft Infrared Countermeasures (LAIRCM) laser turret with two MWS sensors adjacent on the fuselage sides. Other sensors and turrets were installed on the forward quarter of the C-17. These are two of the total six sensors on the aircraft. While such laser systems are principally intended to defeat infrared guided MANPADS, these can be effective against comparable AAM. The system weighed several hundred pounds and consumed significant volume, plus adding electrical load and drag, making it unsuitable for fighter aircraft applications without substantial downsizing. (USAF)

the high-powered laser pulse rate. The high-power engagement laser then targets the missile. If required to cover the entire sphere around the aircraft, several such turrets would be required. This Directional Infrared Countermeasures (DIRCM) is more likely to be installed on transport-type aircraft, at risk from MANPADS, which have the volume and weight margin for such equipment. Correct modulation for the threat system (laser pulse rate) continues to be important. Making this a programmable feature of the system, and adjustable in the field, helps to ensure effectiveness against anticipated adversary threats in the theater of operations. Lasers powerful enough to damage the AAM while in flight were achieved by the 2020s. Missile systems can be 'hardened' to reduce the effectiveness of the DIRCM laser.

The missile infrared counter-countermeasures (IRCCM) has, in fourth and fifth generation systems, come to include more and more microcomputer processing in the tracking logic to detect jamming efforts and reject divergent signals. This has included detecting the abrupt switch to a source ultimately rejected and then return to the initial target via flightpath prediction calculations. Such calculations can also be used to discern the behavior of a flare as not that of an aircraft and so reject it on that criterion. All these factors require IRCM to be continually improved and to increase the J/S to remain effective, and so the IRCCM to likewise improve in the endless cycle. When it comes to an IIR system countering a DIRCM system (DIRCCM), a detector is added to the seeker to sense the intense laser light and it is deflected away from the seeker's IR detector array via a prism.

Any Mode

Efforts were underway from the 1990s to develop airborne laser systems capable of targeting and engaging missiles in flight with sufficient intensity to 'scramble' the electronic or heat up the airframe until it fails under flight loads or causes the motor and warhead to explode. This rested both on the ability to detect and track the missile even following motor burnout, then firing a laser with great precision against a small and rapidly moving target. The laser itself relies upon high power generation and other mechanisms at volumes and weights compatible with tactical aircraft installation – daunting demands that required more than a decade of work. Demonstrations by the USA were favorable, but a fieldable system remained just out of reach.

In Summary

Table 10 summarizes the factors that affect engagements employing AAM. How these weigh in the favor of the shooter or target (and the aircrews' ability to perform that mental assessment) will determine the outcome – and it is seldom one factor that decides success or failure. The targeted pilot may also be going through the shooter factors and attempting to get off a shot before the opposition AAM arrives.

Table 10: Factors Affecting Engagements Employing AAM

Factor	Applicable to		
	Shooter	Target	Both
Aircrew experience			√
Aircrew proficiency			√
Maneuver capability of aircraft			√
All in-use systems maintained and functioning per specifications			√
AAM available	√		
Selection of most suitable AAM	√		
Capabilities of AAM	√		
Available countermeasures		√	
Available counter-countermeasures	√		
Effectiveness of available CM or CCM			√
Off-board guidance to favorable intercept aspect or evasion (GCI, AWACS)			√
Earliest detection of target or shooter (radar, IRST, eyesight)			√
Ability to execute an undetected approach and launch	√		
Maneuvering to advantage for a shot against a maneuvering target	√		
Maneuvering to advantage to spoil shot		√	
First shot opportunities	√		
Decision on volley fire	√		
Time required to ready and launch AAM	√		
Onboard guidance or pilot assessment of best opportunity to shoot	√		
Aircrew detection of launch		√	
Onboard guidance or pilot assessment of best opportunity to deploy CM		√	
Aircrew maneuvers to defeat AAM or minimize potential for lethal intercept		√	
Missiles reaches target	√		
Missile fuzes suitably	√		
Missile warhead detonation destroys target	√		
Unexpected and intangibles			√

Given all the above factors, the potential for a miss is significant. This assumes a one-on-one engagement with only one aircraft firing. A scenario with both jets firing AAM, or a many-on-many battle, would be more complex but involve the same mix of factors.

The following volumes will show how AAM matured over the decades in support of the elements above in which weapon capabilities were vital.

BIBLIOGRAPHY

Cleaver, Thomas McKelvey, *The Tonkin Gulf Yacht Club, Naval Aviation in the Vietnam War* (Oxford, England: Osprey Publishing, 2021)

Dabrowski, Krzysztof, *Defending Rodinu Volume 2: Build-up and Operational History of the Soviet Air Defence Force 1960–1989* (Warwick, England: Helion, 2023)

Eichblatt, Emil J., *Test and Evaluation of the Tactical Missile*, Progress in Astronautics and Aeronautics Volume 119 (Washington, D.C.: American Institute of Aeronautics and Astronautics, 1989)

Etter, James P., 'Air-to-Air Missile Training: Out of Step with Reality', *Marine Corps Gazette*, May 1991, pp.54–55

Forbat, John, *The Secret World of Vickers Guided Weapons* (Gloucestershire, England: The History Press, 2010), e-book

Olds, Robin with Olds, Christina and Rasimus, Ed, *Fighter Pilot: The Memoirs of Legendary Ace Robin Olds* (New York, New York: St. Martin's Press, 2010)

Richardson, D. Kenneth, *Hughes After Howard, The Story of Hughes Aircraft Company* (Santa Barbara, California: Sea Hill Press, 2011)

Shaw, Robert L., *Fighter Combat, Tactics and Maneuvering* (Annapolis, Maryland: Naval Institute Press, 1985)

Walker, John R., *Air Superiority Operations*, Brassey's Air Power Air Power: Aircraft, Weapon Systems and Technology Series Volume 5 (London, England: Brassey's Defence Publishers, 1989)

FURTHER READING

Books

Department of the Air Force, *Guided Missiles Fundamentals*, AFM 52-13 (Washington, D.C.: Headquarters US Air Force, 1 July 1972)

Elsam, M.B., *Air Defence*, Brassey's Air Power Air Power: Aircraft, Weapon Systems and Technology Series Volume 7 (London, England: Brassey's Defence Publishers, 1989)

Fitts, Richard E., ed., *The Strategy of Electromagnetic Conflict* (Los Altos, California: Peninsula Publishing, 1980)

Gunston, Bill, *The Illustrated Encyclopedia of Aircraft Armament* (London, England: Salamander Books, 1987)

Gunston, Bill, *The Illustrated Encyclopedia of the World's Rockets & Missiles* (London, England: Salamander Books, 1979)

Gunston, Bill, *Illustrated Guide to Modern Airborne Missiles* (London, England: Salamander Books, 1983)

Mason, R.A., *Air Power: An Overview of Roles*, Brassey's Air Power Air Power: Aircraft, Weapon Systems and Technology Series Volume 1 (London, England: Brassey's Defence Publishers, 1987)

REPORTS, PAPERS, AND LECTURES

Deuerle, Craig M., 'Reticle Based Missile Seekers', in Driggers, Ronald G. (ed.), *Encyclopedia of Optical Engineering*, Volume 3, (New York New York: Marcel Dekker, 2003), pp. 2400–2408

Fino, Steven A., "All the Missiles Work": Technological Dislocations and Military Innovation, A Case Study in US Air Force Air-to-Air Armament Post-World War II through Operation Rolling Thunder (Maxwell AFB, Alabama: Air University Press. January 2015)

Hpasp, ed., 'History of Electro-Optical Guided Missiles' Version 1.01, 15 October 2016

Kopp, Carlo. 'Active and Semiactive Radar Missile Guidance' (Australian Aviation, 2005), https://www.ausairpower.net/TE-Radar-AAMs.html

Kopp, Carlo. 'Heat-Seeking Missile Guidance' (Australian Aviation, 2005), https://www.ausairpower.net/TE-IR-Guidance.html

Olsson, Gustaf, 'Reticles', in Driggers, Ronald G. (ed.), *Encyclopedia of Optical Engineering*, Volume 3, (New York New York: Marcel Dekker, 2003), pp. 2409–2416

Stillion, John, 'Trends in Air-to-Air Combat, Implications for Future Air Superiority' (Washington, D.C.; Center for Strategic and Budgetary Assessments, 2015)

INTERNET

Anon, 'Air-to-Air Missiles', War Thunder Wiki, https://wiki.warthunder.com/Air-to-air_missiles, accessed 12 March 2024

Anon., 'Air-to-Air Missile', Wikipedia, https://en.wikipedia.org/wiki/Air-to-air_missile, accessed 12 March 2024

ENDNOTES

Chapter 1

1 The acronym IRH is an invention of the author and is not found in common usage within the literature.

2 The definition of WVR is subject to much variability in the literature. From the author's aviation experience, an average aviator with good eyesight under good visual conditions can perceive a fighter-size airborne object at up to 3mi (5km). Making a positive identification of aircraft type is practical at around 1.5mi (2.5km). Beyond visual range would be in excess of three miles.

3 There was an incident in 1956 when radio control of an F6F drone out of Naval Air Station (NAS) Point Mugu, California, was lost and the aircraft wandered aimlessly over the Los Angeles basin. Two F-89 fighters armed with rockets were sent up to shoot down the drone over a safe area before it endangered civilians. None of the 208 rockets fired hit the F6F but did start brushfires and one ended up in the radiator of a pickup truck traveling along a road, fortunately without the warhead detonating. The drone eventually came down when its fuel was exhausted.

4 Extracted and revised from the translation of the MiG-21F-13 and MiG-21PF interceptor tactical manual reproduced in Krzysztof Dabrowski, *Defending Rodinu Volume 2: Build-up and Operational History of the Soviet Air Defence Force 1960–1989* (Warwick, England: Helion, 2023), p.19.

5 An FCS is an aircraft system that uses outputs of several aircraft systems, such as radar and navigation, to ensure the missile is provided data (if so designed) and the pilot presented information to assist in firing the AAM with the greatest potential for intercepting the intended target.

6 g-level refers to the perceived force felt as a body traveling in a straight line is disturbed by turns in the lateral axis or push-over and pull-ups in the vertical, called centripetal acceleration. This is measured as multiples of the acceleration due to gravity (32.2fps/9.8mps in freefall) or 1g. The vehicle's structure must be designed to withstand the stresses from such maneuvering.

7 Mach is a measure of the speed of sound at the operating altitude. Thus, a Mach number has different airspeed by other measures, such as speed over the ground, at different altitudes owing to varying air density.

8 Load refers to the forces applied to a structure that must be reacted to prevent structural failure or excessive deformation. These loads can be aerodynamic pressure distribution (airloads), forces imparted by the aircraft launcher and the AAM's motor, the load factor (inertial forces), and more.

9 Slant range is the distance a missile travels from launch to intercept rather than the horizontal distance measured across the ground.

10 Handling qualities refers to ease of pilot operation of the flight controls with precision in executing the intended mission. Flying qualities refers to aircraft response to pilot inputs and air gusts, with both static and dynamic stability playing a role. All may be adversely effected by carriage and launch of an AAM with consequent bearing on flight safety.

11 Air-to-air missiles have no commercial applications and so are developed and produced by nation states. Some development is begun by private or public defense companies as R&D efforts with expectation they may eventually be taken up by the state. No significant production is undertaken in such circumstances owing to excessive outlay without certainty of return on investment.

Chapter 2

1 By the 1990s the military customer was, in many product fields, unable to compete with commercial market pressures. Their contracts were too cumbersome to manage, and production runs too shallow, regardless of the money offered, for the manufacturers to turn design and production capacity away from more lucrative commercial markets. The only alternative was to create an in-house production or reconditioning capability at depots or to develop an entirely new system on a contract attractive to defense manufacturers.

2 Emil J. Eichblatt, *Test and Evaluation of the Tactical Missile*, Progress in Astronautics and Aeronautics Volume 119 (Washington, D.C.: American Institute of Aeronautics and Astronautics, 1989), p.170, with modifications.

3 An inertial navigation system uses motion sensing devices like accelerometers, gyroscopes and altimeter, plus a timer, to determine the missile's orientation, position and velocity in space from a start location and without external references.

4 D. Kenneth Richardson, *Hughes After Howard, The Story of Hughes Aircraft Company* (Santa Barbara, California: Sea Hill Press, 2011), pp.338–339. The event was a shot of an AIM-7E-2 off F-14A 157985 on 20 June 1973 flying out of NAS Point Mugu at 5,000ft (1,525m) and 700mph (1,125kph).

Chapter 3

1 The 'gains' referred to are sensitivity of a function adjustment, like a hot water valve – a small adjustment producing a large change in water temperature is high gain while a large adjustment with small change is low gain.

2 Eichblatt, *Test and Evaluation of the Tactical Missile*, p.352, with modifications.

3 Eichblatt, *Test and Evaluation of the Tactical Missile*, p.350, with modifications.

4 Eichblatt, *Test and Evaluation of the Tactical Missile*, p.352, with modifications.

5 Eichblatt, *Test and Evaluation of the Tactical Missile*, p.55, with modifications.

6 Negative g is seldom employed in air-to-air combat, but fighters were initially capable of maneuvering to -1 and later to -3g. The typical axisymmetric missiles, without aircrew, are not challenged in exceeding such load factors.

7 A ramjet engine uses an inlet, accelerated to high-speed, to compress incoming air for combustion and so eliminates rotating compressor and turbine components common in a jet engine. The speed at which the ramjet inlet functions suitably must be achieved by some other means as the ramjet lacks low-speed thrust. Once operating, thrust increases with airspeed and so the missile continually accelerates until fuel exhaustion. The areas of the inlet diffuser (decreasing airflow velocity while increasing pressure), combustion section, and nozzle exit area must be carefully matched to ensure efficient operation.

8 Bleeding refers to collecting and routing the high-pressure gas, generated by combustion, via one or more tubes.

9 Global Positioning System is a satellite-based navigation system that can fix the location of a transceiver to within inches. The USA first placed GPS into operation and other nations or coalitions followed with their own systems.

10 The literature frequently adds the word 'active' before radar or RF and laser fuze references. This book, in its six volumes, will not do this as it is redundant.

11 Eichblatt, *Test and Evaluation of the Tactical Missile*, p.66, with changes.

12 This radar has the ability to differentiate returned ground clutter (confused returns) and objects moving across that background by differences in frequency of the returns via the Doppler Effect. This offers a 'look-down/shoot-down' capability. Where this difference is small, the range data may be poor.

13 The monopulse uses multiple offset lobes of the beam instead of a single beam in a typical conical scan system and so has fewer confused returns. Returns from the multiple lobes are processed for target data of higher accuracy.

14 In radar application (microwaves) the waveguides are typically metal pipes of rectangular cross-section that conduct the energy over short distances with minimal loss. (Microwaves cannot be conducted via cables.)

15 Eichblatt, *Test and Evaluation of the Tactical Missile*, p.353, with modifications.

Chapter 4

1 John Forbat, *The Secret World of Vickers Guided Weapons* (Gloucestershire, England: The History Press, 2010), e-book, unpaginated (search for 'inadvertently ignited inside').

2 Robin Olds with Christina Olds and Ed Rasimus, *Fighter Pilot: The Memoirs of Legendary Ace Robin Olds* (New York, New York: St. Martin's Press, 2010), p.305.

3 James P. Etter, 'Air-to-Air Missile Training: Out of Step with Reality', *Marine Corps Gazette*, May 1991, pp.54–55. This changed rapidly after the date of publication.

4 A Chinese pilot reported watching one of two SARH AIM-7D Sparrows fired at his flight lead by an F-4B, in a vertical maneuver, miss the intended target only to have one of the rounds hit the F-4 which then crashed. Thomas McKelvey Cleaver, *The Tonkin Gulf Yacht Club: Naval Aviation in the Vietnam War* (Oxford, England: Osprey Publishing, 2021), e-book, unpaginated (search for 'hit by one of the malfunctioning missiles'). In this 9 April 1964 engagement between US Navy Phantoms and Chinese MiG-17s, one Sparrow went ballistic (flew down range without tracking), and three failed to ignite to just fall away when ejected. A GCI radar plot showed the Navy jet being followed by another aircraft and eventually disappearing from the screen, presumably having been shot down. So, the Chinese officer's account, related in 1994, has been considered either an erroneous recollection or a fabrication.

5 These are re-drawings of those depicted in John R. Walker, *Air Superiority Operations*, Brassey's Air Power Air Power: Aircraft, Weapon Systems and Technology Series Volume 5 (London, England: Brassey's Defence Publishers, 1989), pp.56 and 58.

6 These are re-drawings of those depicted in Robert L. Shaw, *Fighter Combat, Tactics and Maneuvering* (Annapolis, Maryland: Naval Institute Press, 1985), p.47. Shaw attributes these to a Doppler radar homing AAM but is representative for any missile in most respects.

7 Cleaver, *The Tonkin Gulf Yacht Club*, e-book, unpaginated (search for 'one of his missiles caught fire').

8 The f-pole is the maximum horizontal component of separation between the missile and the launch aircraft during the fly-out. Systems with large f-pole capability suggests engagement from a farther distance, typically advantageous. A-pole is the instance when the round begins active tracking.

9 A basic radar reflector is the corner reflector, a simple assembly of radar-reflective material, such as aluminum, creating an interior corner. The most common is the inside of a three-side pyramid. The incident radar energy bounces about inside the reflector until reflected back out. The reflector is installed with the opening facing to the rear and generally behind an RF-permeable fairing. A more advanced type is the Luneburg Lens which is a spherical reflector, returning RF signals originating from any bearing and elevation.

ABOUT THE AUTHOR

William J. Norton is retired from a flight test engineering career that spanned 40 years, including 20 as a US Air Force officer. He has held numerous positions in many organizations on dozens of aerospace programs spanning all aircraft types. He has penned scores of technical papers, 17 books, and a multitude of magazine articles. Bill holds a Masters in Aeronautical Engineering and has taught courses at the college-level. He is a civil pilot with numerous ratings, restored and operated a DHC-1 Chipmunk, and built and flight-tested a Rutan Long-EZ. Bill is married to the lovely and talented Anya Victoria Eriksson.